OXFORD
UNIVERSITY PRESS

ESSENTIAL
CHEMISTRY STAGE 9
FOR CAMBRIDGE SECONDARY 1

Lawrie Ryan

Roger Norris | Editor: Lawrie Ryan

Oxford excellence for Cambridge Secondary 1

OXFORD

Great Clarendon Street, Oxford, OX2 6DP, United Kingdom

Oxford University Press is a department of the University of Oxford.
It furthers the University's objective of excellence in research, scholarship,
and education by publishing worldwide. Oxford is a registered trade mark of
Oxford University Press in the UK and in certain other countries

First published by Nelson Thornes Ltd in 2013
This edition published by Oxford University Press in 2015

British Library Cataloguing in Publication Data
Data available

9780198399896

10 9 8 7 6 5 4 3

Printed in Great Britain by CPI Group (UK) Ltd., Croydon CR0 4YY

Acknowledgements

Cover photograph: © Getty Images/Doris Rudd Designs, Photography
Illustrations: Tech-Set Ltd
Page make-up: Tech-Set Ltd, Gateshead

Although we have made every effort to trace and contact all
copyright holders before publication this has not been possible in all
cases. If notified, the publisher will rectify any errors or omissions at
the earliest opportunity.

Links to third party websites are provided by Oxford in good faith
and for information only. Oxford disclaims any responsibility for
the materials contained in any third party website referenced in
this work.

Contents

Access your support website:
www.oxfordsecondary.com/9780198399896

Welcome to *Science for Cambridge Secondary 1!*
This student book covers Stage 9 Chemistry of the curriculum and will ensure you are prepared for your IGCSE in Chemistry.

Using this book

This book covers one of the main disciplines of science, Chemistry, though you will find some overlap with the other two subject areas, Biology and Physics. Each chapter starts with *Science in context!* pages. These pages put the chapter into a real-world context, and provide a thought-provoking introduction to the topics. You do not need to learn or memorise the information and facts on these pages; they are given for your interest only. Key points summarise the main content of the chapter.

The chapters are divided into topics, each one on a double-page spread. Each topic starts with a list of learning outcomes.

These tell you what you should be able to do by the end of that topic.

Learning outcomes

Key terms are highlighted in **bold type** within the text and definitions are given in the glossary at the end of the book. Each topic ends with a summary of the key terms you should understand and remember.

Key terms

Summary questions at the end of each topic allow you to assess your comprehension before you move on to the next topic.

Summary questions

Expert tips are used throughout the book to help you avoid any common errors and misconceptions.

Expert tips

Practical activities are suggested throughout the book, and will help you to plan investigations, record your results, draw conclusions, use secondary sources and evaluate the data collected.

Practical activity

At the end of each chapter there is a double page of examination-style questions for you to practise your examination technique and evaluate your learning so far.

Answers to summary questions and end of chapter questions are supplied on a separate Teacher's CD.

Student's website

The website included with this book gives you additional learning and revision resources in the form of interactive exercises, to support you through Stage 9 Chemistry.

www.oxfordsecondary.com/9780198399896

Science *in context!*

Sodium fluoride – an important salt

You might know that many toothpastes now contain fluoride. The fluoride is usually sodium fluoride, NaF. It is added to protect your teeth from decay. But some water companies also add fluoride to the water supply as well. So if you drink this tap water you also get a small dose of fluoride.

There is some debate about whether fluoride should be added to public water supplies. There are lots of arguments for and against it. The evidence is collected by comparing a group of people who do take fluoride with a group who do not. This means that arguments are often based on research that is open to criticism because the research doesn't control the different variables well enough.

Brushing your teeth is an important part of personal hygiene

Here are some of the arguments for and against fluoridation:

Arguments FOR fluoridation of water	Arguments AGAINST fluoridation of water
Some areas have had fluoridated water for about 50 years now and nobody has proved that there are any harmful effects. The effects of fluoridation in the latest studies show about a 30 per cent reduction in cavities in teeth. This result is not as good as those shown by studies in the 1960s. The old studies showed you were 5 times more likely to have tooth decay if your water was not fluoridated. However, this could be because the bacteria that cause tooth decay are dying out because of the success of fluoridation. We need fluoridation to protect the teeth of those people who do not have good dental hygiene habits and who don't visit their dentist regularly. The bacteria associated with tooth decay also cause some types of heart disease, so fluoridation will protect us from that. The fluoride is only added in tiny amounts (1 part per million), which is well below harmful levels.	What happens to your teeth reflects what's happening to your bones. Fluorosis is a condition caused when children take too much fluoride. White streaks or tips appear on their teeth. These are harmless deposits of calcium fluoride. They are porous and can become stained. Fluorosis could be a sign of other changes in your bones. Some studies have linked **excess** fluoride to weakening of bones (increasing numbers of fractures) and bone cancer. The benefit of fluoridation for teeth is not significant (accounting for less than one filling saved per person). So why should we take any risks with our health? Toothpastes and dental care have improved since the 1960s. The claims of huge benefits of fluoridation were not really proved then, so we don't need it now. It is wrong to give people treatments that they have not consented to. People have a right to choose. Some studies show that **excess** fluoride affects the brain, producing learning difficulties. It has also been associated with Alzheimer's disease in old people. You can't set safe limits of fluoride because you can't control how much people take. Children, especially, might swallow more toothpaste than adults – look at the warnings on the tubes!

Researchers find it difficult to set up a 'fair test' to compare areas where the water is fluoridated (has sodium fluoride added) with areas where no fluoride is added. There are many other factors that may vary between the people living in these different test areas. For example, the average age, income, diet or dental care available may differ. So increasing the size of the samples and matching similar groups to compare them is the best scientists can do to get data that is as vaild as possible.

There are many such debates on issues that affect us all. Scientists can gather evidence to help inform the debate but society has to ultimately decide what developments are made, often through government policies. In this chapter you will find out about the structure of an atom and its discovery by scientists. You will also learn more about the Periodic Table and the preparation of some common salts by the reactions of metals or metal carbonates with acid.

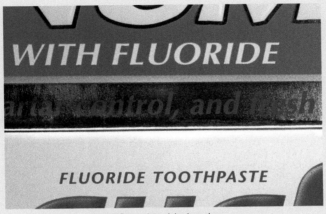
Many toothpastes have fluoride added to them

Key points

- Atoms contain protons, neutrons and electrons.
- Protons are positively charged. Electrons are negatively charged. Neutrons have no charge – they are neutral.
- Protons and neutrons are the heavy particles in an atom.
  They are found in the nucleus (centre) of an atom. (We can ignore the tiny mass of the electrons in an atom.)
- The electrons orbit the nucleus in shells (or energy levels).
- The 1st shell can hold 2 electrons.
  The 2nd shell can hold 8 electrons, as can the 3rd shell before the 4th shell starts filling.
- The atomic number is the number of protons (which equals the number of electrons) in an atom. It also tells us the position of an element in the periodic table.
- The periodic table arranges the elements in order of atomic number.
- Elements with similar properties line up in vertical columns. These columns are called groups.
- There are 8 main groups in the periodic table.
- A row across the periodic table is called a period.
- The elements can be divided into metals and non-metals (with a few semi-metals or metalloids in between).
- There are trends in properties going down groups and across periods in the periodic table.
- **carbonate + acid → a salt + water + carbon dioxide**

Revisiting the Periodic Table

You already know that the chemical elements are organised into a special order in the **Periodic Table**. You also learned the symbols of the atoms of the first twenty elements. Can you cover the 2nd and 4th columns in the table below and still recall the chemical symbol of each atom?

Atom	Symbol	Atom	Symbol
hydrogen	H	sodium	Na
helium	He	magnesium	Mg
lithium	Li	aluminium	Al
beryllium	Be	silicon	Si
boron	B	phosphorus	P
carbon	C	sulfur	S
nitrogen	N	chlorine	Cl
oxygen	O	argon	Ar
fluorine	F	potassium	K
neon	Ne	calcium	Ca

In Stage 8, you did work on the metals and non-metals in the Periodic Table. You also found the state (solid, liquid or gas) of elements at room temperature. This is summarised in the following tables:

Element	Metal or non-metal	Element	Metal or non-metal
hydrogen	non-metal	sodium	metal
helium	non-metal	magnesium	metal
lithium	metal	aluminium	metal
beryllium	metal	silicon	Semi-metal or metalloid, with many reactions like a non-metallic element.
boron	non-metal	phosphorus	non-metal
carbon	non-metal	sulfur	non-metal
nitrogen	non-metal	chlorine	non-metal
oxygen	non-metal	argon	non-metal
fluorine	non-metal	potassium	metal
neon	non-metal	calcium	metal

Element	State at room temperature (20°C)	Element	State at room temperature (20°C)
hydrogen	gas	sodium	solid
helium	gas	magnesium	solid
lithium	solid	aluminium	solid
beryllium	solid	silicon	solid
boron	solid	phosphorus	solid
carbon	solid	sulfur	solid
nitrogen	gas	chlorine	gas
oxygen	gas	argon	gas
fluorine	gas	potassium	solid
neon	gas	calcium	solid

Expert tips

The noble gases, in Group 0 of the Periodic Table, had not been discovered in 1869 when Dmitri Mendeleev first proposed his table. They were added later as each one was discovered.

Remember that a Russian chemist called Dmitri Mendeleev first proposed the structure of the modern Periodic Table. He put the elements in order of their atomic mass, starting new lines so that similar elements lined up with each other.

The first twenty elements in the Periodic Table are arranged like this:

H 1							He 2
Li 3	Be 4		B 5	C 6	N 7	O 8	F 9	Ne 10
Na 11	Mg 12		Al 13	Si 14	P 15	S 16	Cl 17	Ar 18
K 19	Ca 20						

The first twenty elements of the Periodic Table

Summary questions

1 In the first twenty elements of the Periodic Table name the:
   a) metallic element with the largest atomic mass
   b) element with the lightest atoms
   c) semi-metal (metalloid).

2 a) Who was the scientist who first proposed the Periodic Table?
   b) What country was this scientist from?

3 Choose one element from the first twenty elements of the Periodic Table and find out its:
   a) melting point
   b) boiling point
   c) density
   d) relative atomic mass.

You have an idea what atoms are like from your work in Stage 8. The atoms were described as tiny, hard spheres, with each element having atoms of different sizes and masses. Scientists used this model until near the end of the 19th century. At that time, great advances were made as scientists gathered evidence about what atoms themselves are made of.

J.J. Thomson

In 1897, a scientist called J.J. Thomson discovered the **electron**. This is a tiny negatively charged particle that is much, much smaller than any atom. When he discovered the electron, Thomson was experimenting with a sealed glass tube containing gas at low pressure. He applied very high voltages to his apparatus.

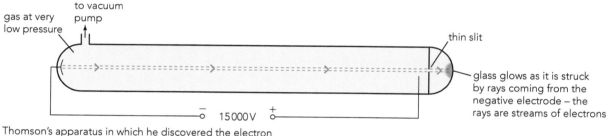

Thomson's apparatus in which he discovered the electron

He noticed an interesting effect. A beam of particles that made glass at the end of the tube glow was detected. He named these particles electrons.

Thomson did experiments on the beams of electrons in the glass tube. They were attracted to a positive charge, so he concluded that they must be negatively charged. Other experiments showed that about 2000 electrons had the same mass as the lightest atom, hydrogen.

But where had these tiny electrons come from? Since they were so small, Thomson suggested that they could only have come from inside atoms. So Thomson proposed a new model for the atom.

He said that the tiny negatively charged electrons must be within a cloud of positive charge. Thomson imagined the electrons as the bits of plum in a plum pudding (rather like currants spread through a Christmas pudding – but with lots more space in between).

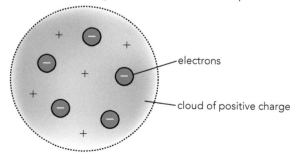

Thomson's 'plum pudding' model of the atom

Ernest Rutherford

The next development came about 15 years later. Two of Ernest Rutherford's students were doing an experiment with radiation. The radiation consisted of dense, positively charged particles called alpha particles. These were used like 'atomic bullets', to fire at a very thin piece of gold foil.

They predicted the particles would bash their way straight through the gold atoms. Rutherford and his students based the prediction on the plum pudding model of the atom. They thought that the cloud of positive charge and tiny electrons spread throughout the atom could not stop the alpha particles. However, they got a big surprise.

Look at their experiment below:

Expert tips

In the early 20th century, scientists were able to carry out experiments, not possible in earlier times. They explained their new observations by developing the model of the atom so that the observations made sense. Scientists are still exploring the structure of the atom today and coming up with new ideas.

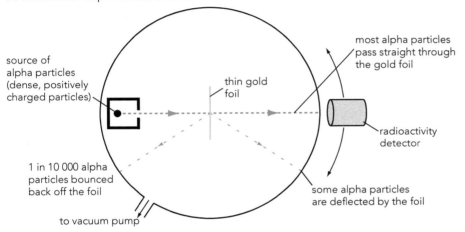

Rutherford's experiment – the gold atoms had somehow deflected and even stopped some alpha particles from passing through them

In 1911, Ernest Rutherford interpreted these results. He suggested a new model for the atom. He said that Thomson's plum pudding model could not be right. He proposed that the positive charge in an atom must be concentrated in a tiny volume at the centre of the atom. Otherwise the alpha particles fired at the foil could never be repelled back from the atoms in the gold foil.

In Rutherford's new model of the atom, the negative electrons orbited around a small, dense **nucleus** (centre of the atom).

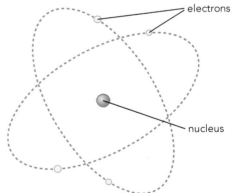

Rutherford's nuclear model of the atom

Summary questions

1 Which type of particle did J.J. Thomson discover?

2 Describe Thomson's plum pudding model of the atom.

3 a) Why was Rutherford's model of the atom called the 'nuclear model' of the atom?

   b) Describe the experiment that led to Rutherford changing the plum pudding model of the atom. Explain his reasoning.

The next important development in the structure of the atom came in 1914. Danish physicist Niels Bohr revised the model again. This time Rutherford's model (see page 7) had to be changed to explain further observations.

It had been known for some time that light was given out when atoms were heated. The light always had particular amounts of energy. But, as yet, no one had been able to explain this. Then Bohr suggested that the electrons must be orbiting the nucleus in certain fixed **energy levels** (or **shells**).

The energy must be given out when 'excited' electrons fall from a higher energy level (shell) to a lower one.

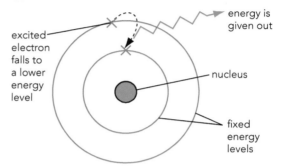

Bohr's model of the atom, with its electron shells (or energy levels)

The first electron shell, nearest the nucleus, can hold a maximum of two electrons. The second shell can hold maximum of eight electrons. The next electrons enter the third shell. When the third shell holds eight electrons the next two electrons occupy the fourth shell.

The arrangement of electrons in atoms

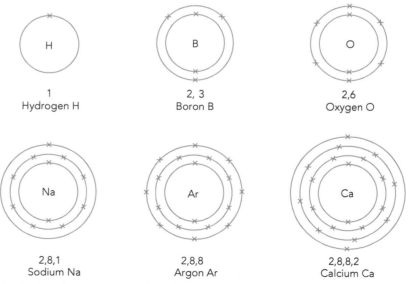

| 1 | 2, 3 | 2,6 |
| Hydrogen H | Boron B | Oxygen O |

| 2,8,1 | 2,8,8 | 2,8,8,2 |
| Sodium Na | Argon Ar | Calcium Ca |

A selection of atoms, showing their electron arrangements

What's in the nucleus?

All atoms are neutral – they have no overall electrical charge on them. So if we know the number of electrons (negatively charged) we know this is the same as the number of **protons** (positively charged).

The protons are found in the nucleus, along with sub-atomic particles with the same mass called **neutrons**. The neutrons have no charge. They are neutral.

So an atom of carbon, with six protons, six electrons and six neutrons can be thought of like this:

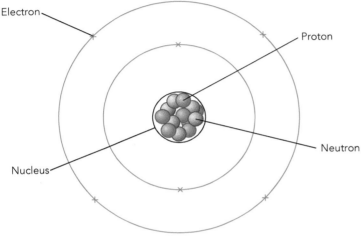

A carbon atom

Key terms

- **energy level (shell)**
- **neutron**
- **proton**

Summary questions

1. In which part of an atom do we find:
   a) protons
   b) electrons
   c) neutrons?

2. Where is the mass of an atom concentrated?

3. Draw an atom of boron (B), showing the number and location of its protons, neutrons and electrons. (B has five protons, five electrons and six neutrons).

The alkali metals are stored in oil because they are so reactive

Elements in the same group of the Periodic Table are similar, but not the same. They often show **trends** in their properties within a group.

Group 1

The elements in Group 1 are all reactive metals. Your teacher will show you some trends in the properties of Group 1 elements. The first three metals in Group 1 are:

Li – lithium   Na – sodium   K – potassium.

Practical activity   Trends in Group 1

Watch your teacher cut a small piece of lithium from a lump of the metal using a knife.

Compare this with the ease of cutting sodium metal with the same knife.

- Which metal is easier to cut, lithium or sodium?
- Predict how easy it will be for your teacher to cut a piece of potassium.

Your teacher will test your prediction.

- What is the trend in the hardness of the Group 1 metals going down the group?

Watch your teacher add a small piece of lithium and the sodium to a trough of water.

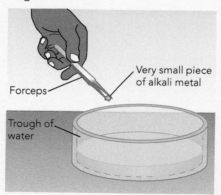

- Compare the two reactions. Which is more reactive, lithium or sodium?
- Predict the reaction of potassium with water.
- Watch your teacher carry out the reaction. Was your prediction correct?

⚠ **Watch these demonstrations from behind a safety screen and wear eye protection.**

We find that the Group 1 elements get softer, and get more reactive, going down the group.

Group 7

The elements in Group 7 are all non-metals:

F – fluorine

Cl – chlorine

Br – bromine

I – iodine.

chlorine

bromine

iodine

The elements in Group 7 are non-metals unlike the metals in Group 1 at the opposite side of the Periodic Table

 IGCSE Link...
You will lean more about different groups in the Periodic Table in IGCSE Chemistry.

Practical activity   Looking for patterns

Look at the table below:

Halogen molecule	Colour	State (at 25 °C)
F_2	pale yellow	gas
Cl_2	yellow/green	gas
Br_2	orange/brown	liquid
I_2	grey/black (violet vapour)	solid

The Group 7 elements all form 'two-atom' molecules, such as F_2 and Cl_2.

You can also see some patterns in the table going down the group.

• Do they get darker or lighter in colour?

Look at their states:

• What is the trend?
• Do their melting points and boiling points get higher or lower as we go down the group?
• Using the table, make some predictions about astatine (At), the element at the bottom of Group 7, below iodine.

Summary questions

1 What is the trend in reactivity going down Group 1?

2 What is the trend in the colour of the elements going down Group 7?

3 Rubidium (Rb) lies beneath potassium (K) in the Periodic Table.
   a) What group is rubidium in?
   b) Predict how difficult it would be to cut a piece of rubidium with a knife.
   c) Predict what you would see when a piece of rubidium is dropped into a trough of water.

Atomic structure and the Periodic Table

You need to be able to draw an atom of any of the first twenty elements in the Periodic Table. You must show the arrangement of the electrons in the shells (energy levels) around the nucleus.

The atoms of each element in the Periodic Table have one more proton (and electron) than the one before it. The elements are arranged in order of their **atomic number**. This is also known as the proton number. The atomic number of an element is defined as the number of protons in each of its atoms. This number also equals the number of electrons in each of its atoms. Remember that atoms are neutral. So each atom always has the same number of protons and electrons.

The atomic number of each element is usually shown at the bottom left of its box in the Periodic Table.

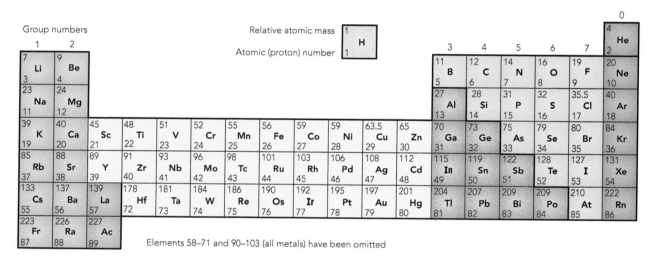

Elements 58–71 and 90–103 (all metals) have been omitted

Look at the atoms below, taken from the first two and last two groups of the Periodic Table:

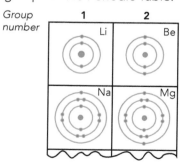
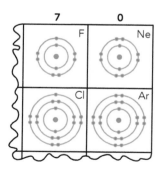

- Can you see a link between the group number and the number of electrons in the outer shell?

We find that **the number of electrons in the outer shell equals the group number**.

There is also a link between an element's period number and its atomic structure. The periods are the horizontal rows across each line of the Periodic Table.

- The 1st period has just two elements (H and He).
- The 2nd period, starting with Li and finishing with Ne, has eight elements.
- The 3rd period, starting with Na and finishing with Ar, also has eight elements.
- The 4th period, starting with K and Ca – the 19th and 20th elements, has 18 elements.

Draw one atom in Group 2 from the 2nd, 3rd and 4th periods.

Can you see a link between the number of electron shells and the period?

The period number equals the number of occupied electron shells in an atom.

Notice that a new electron shell (energy level) starts to be filled as each period starts.

Expert tips

When asked to draw an atom, showing the arrangement of its electrons, you will be given information to tell you the number of electrons in the atom (equal to the atomic number). You do not need to memorise an element's atomic number.

Key terms

- **atomic number**

Summary questions

1 Define 'atomic number'.
2 What is the link between the group number of an element and the arrangement of its electrons?
3 Iodine (**I**) has an atomic number of 53.
   a) What information does this give us about iodine?
   b) Find iodine in the Periodic Table.
      i What group and period is iodine in?
      ii How many electrons does an atom of iodine have in its outer shell?
      iii How many electron shells are occupied in an iodine atom?

Learning outcomes

After this topic you should be able to:

- write the word equation for a carbonate reacting with an acid
- plan which measurements and observations are necessary and what equipment to use when investigating the reactions of carbonates with acids
- safely prepare crystals of a salt from the reaction of an insoluble carbonate and dilute acid.

Calcium carbonate is attacked by acids

The most common **carbonate** is calcium carbonate, $CaCO_3$. It is the main compound in limestone rock. Some buildings and statues are made from limestone. These structures suffer damage from acid rain. In the next investigation you can find out about the reaction between carbonates and dilute acids.

Practical activity   Investigating metal carbonates and acids

You are given the following apparatus and solutions:

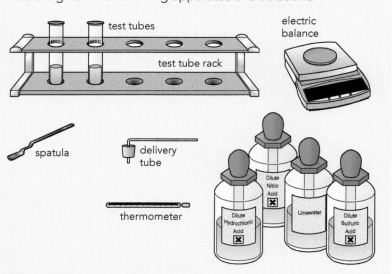

You also have a range of different metal carbonates including copper carbonate, sodium carbonate, potassium carbonate and magnesium carbonate.

Your task is to explore what happens in reactions between the carbonates and the acids. You should record similarities and differences systematically.

Make sure you plan to measure any temperature changes that take place.

⚠ **Let your teacher check your plans before you try them out.**

The general equation for the reaction between a carbonate and dilute acid is:

carbonate + acid → a salt + water + carbon dioxide

For example:

calcium carbonate + hydrochloric acid
→ calcium chloride + water + carbon dioxide

We can use the reaction between a carbonate and an acid to make salts. In the following experiment you will use copper carbonate and dilute hydrochloric acid.

Practical activity   Preparing a salt from an acid and an insoluble carbonate

 Eye protection must be worn.

1 Collect 25 cm³ of hydrochloric acid in a small beaker.

2 Now add a spatula of copper carbonate powder.
   - What happens? What is the gas given off?

3 Add more copper carbonate until it stops fizzing and some solid remains in the beaker.

4 Filter to remove the un-reacted copper carbonate.

5 Pour your solution into an evaporating dish. Then heat it on a water bath as shown:

   Stop heating as soon as you see some crystals around the edge of dish, just above the solution.

6 Leave the solution for a few days to crystallise slowly. If there is still some solution left you can filter the crystals of the salt off.
   - What is the chemical name of your salt?
   - What colour and shape are the crystals prepared in this experiment?

   In the experiment above, the reaction to make the salt is:

   copper carbonate + hydrochloric acid
   → copper chloride + water + carbon dioxide

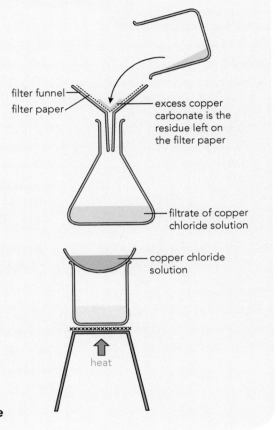

Expert tips

The slower the water evaporates from a salt solution, the larger the salt crystals that can form.

Key terms

- **carbonate**

Summary questions

1 Write down the general equation for the reaction between a carbonate and an acid.

2 Give two ways in which you can tell when all the acid has reacted as copper carbonate (an insoluble carbonate) is added to dilute sulfuric acid.

3 Write down the word equation for the following reactions:
   a) magnesium carbonate + sulfuric acid
   b) nickel carbonate + hydrochloric acid.

Preparing a salt from a soluble carbonate

You have seen how we can use insoluble carbonates, such as copper carbonate, to make salts (see page 15). Can you recall how you can tell when all the acid has reacted in the preparation of the salt? The carbonates give off bubbles of carbon dioxide when they react with acid. So when no more bubbles of gas can be seen the reaction has finished.

To make sure all the acid has been used up, we add the insoluble carbonate in excess. The excess remains as a solid in the solution of the salt formed. We can then filter off the excess carbonate to leave the salt solution as the filtrate.

However, a few carbonates are soluble in water. For these, the method described above does not work. If we add excess soluble carbonate to the acid, the excess dissolves in the solution. There is no solid to filter off. So if we evaporate the water from the solution we would not be left with just salt crystals. We would get a mixture of the salt **and** the un-reacted soluble carbonate.

In the experiment below, you will find when the reaction has just finished by using an indicator. Charcoal is used to remove the coloured dye in the indicator from the salt solution by boiling. The charcoal is then filtered off and the salt can be crystallised from its solution as usual.

The reaction will be between sodium carbonate and dilute hydrochloric acid can be shown by this equation:

sodium carbonate + hydrochloric acid

↓

sodium chloride + water + carbon dioxide

Practical activity   Sodium carbonate plus hydrochloric acid

 Eye protection must be worn.

1 Collect 25 cm³ of hydrochloric acid in a small beaker.

2 Add a few drops of methyl red indicator to the acid.

3 Now add sodium carbonate powder, a small spatula measure at a time, until the fizzing stops and the indicator changes colour.

4 Then add a spatula of charcoal powder to the orange/yellow solution and gently boil. This will remove the colour from the solution.

5 Allow to cool then filter the mixture. You should now have a colourless solution in the conical flask.

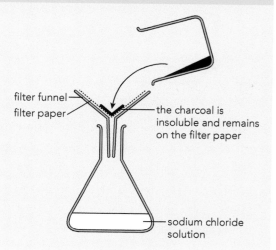

filter funnel
filter paper
the charcoal is insoluble and remains on the filter paper
sodium chloride solution

6 Put the solution into an evaporating dish and heat it on a water bath as shown:

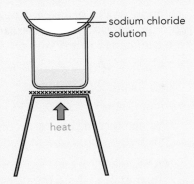

sodium chloride solution

heat

7 Stop heating when you see some white crystals around the edge of the solution.

8 Leave your evaporating dish for a few days. The rest of the water will evaporate off slowly. If there is still some solution left you can filter the crystals of sodium chloride off.
- Write the word equation for the reaction used to make sodium chloride in this experiment.
- What is the shape of the sodium chloride crystals produced in Step 8?

Practical activity   Planning to prepare crystals of zinc nitrate

You are provided with a sample of zinc carbonate powder.

Your task is to plan a detailed method to produce a sample of zinc nitrate crystals.

Start by writing a word equation for the reaction you will carry out. Include diagrams in your method where necessary.

 Do not start any practical work until your teacher has checked your plan.

Summary questions

1 Write down the word equation for the reaction of potassium carbonate and dilute hydrochloric acid.

2 When making crystals of copper chloride using copper carbonate and hydrochloric acid, no indicator is needed in the method. However, when preparing potassium chloride crystals from potassium carbonate and hydrochloric acid, an indicator is needed. Explain this difference between the two methods.

3 Find out what the word 'effervesce' means and use it in a sentence to describe one of the reactions in this chapter.

Throughout this course you have met several reactions of acids that can be used to prepare samples of salts. The general reactions are summarised here, with an example of each type:

1. **metal + acid → a salt + hydrogen**

   For example:

   magnesium + sulfuric acid → magnesium sulfate + hydrogen

   We can test the gas given off with a lighted splint. The hydrogen gas burns with a squeaky pop.

   Note:
   The metals that are unreactive, such as copper and those even less reactive, cannot be used to make their salts using this reaction. They do not react with acid. Also, very reactive metals, such as sodium and potassium, cannot be used. They are explosive in acidic solutions.

2. **metal (base) oxide + acid → a salt + water**

   For example:

   copper oxide + nitric acid → copper nitrate + water

3. **metal hydroxide + acid → a salt + water**

   Some metal hydroxides are soluble in water. These bases are called alkalis.

   For example:

   sodium hydroxide + hydrochloric acid → sodium chloride + water

4. **metal carbonate + acid → a salt + water + carbon dioxide**

   For example:

   calcium carbonate + hydrochloric acid
   ↓
   calcium chloride + water + carbon dioxide

   We can test the carbon dioxide gas given off by bubbling it through limewater. For example:

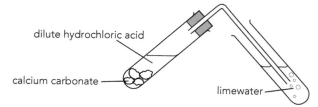

   dilute hydrochloric acid

   calcium carbonate

   limewater

   The limewater turns cloudy white.

5. **Precipitation reactions**

   For example:

   sodium chloride + lead nitrate → lead chloride + sodium nitrate
      (solution)   (solution)   (precipitate)   (solution)

   Precipitation reactions are used to prepare samples of insoluble salts, such as lead chloride, from the solutions of two soluble salts.

Practical activity	Choose your own salt

Choose an example of a salt to prepare from the list below:

magnesium chloride

copper sulfate

calcium nitrate

potassium sulfate

sodium nitrate

calcium sulfate (an insoluble salt)

You can use any of the apparatus commonly found in the laboratory.
You must check the safety information on any reactants you plan to use and on any products made.

- Think about:
  - Which reaction will you use to make the salt?
  - When the reaction involves an acid, what will you observe when all the acid has reacted?
  - What will you do to separate any unreacted solids if necessary?
  - What will you do to make large crystals of a salt?
  - What precautions will you take to make your method as safe as possible?

- Write down your plan in note form, including safety precautions.

 Before starting any practical work, you must check your plans with your teacher.

- Having made your salt, write down an account of your experiment. Include any changes to your original plan. Write why you had to make them and whether they were effective in producing a good example of your chosen salt.

Expert tips

Remember that when you get crystals from a salt solution, you evaporate the water from the solution. Don't say that you evaporate the solution!

Summary questions

1 Copy and complete these word equations:
  a) sodium hydroxide + _____ → sodium chloride + water
  b) potassium carbonate + sulfuric acid → _____ + water + _____

2 Write down the word equation for the reaction of zinc and dilute sulfuric acid.

3 Look at the incomplete word equation below:

_____ + nitric acid → copper nitrate + water

Name TWO compounds that could be the missing compound in the equation.

1 Choose words from the list below to fill in the blanks **a** to **i**.

> **protons** (use twice), **neutrons, eight, energy, centre, two, electrons** (use twice)

The **a** _____ of an atom is called its nucleus.

In the nucleus we find **b** _____ and
c _____

The **d** _____ orbit around the nucleus in
e _____ levels or shells.

The 1st shell can hold a maximum of
f _____ electrons and the 2nd shell can hold
g _____ electrons.

The atomic number of an atom gives us the number of **h** _____, which equals the number of **i** _____ in an atom.   [9]

2 Draw diagrams of these atoms to show the arrangement of their electrons:

  a hydrogen (which has one electron)   [1]

  b boron (which has five electrons)   [1]

  c neon (which has 10 electrons)   [1]

  d magnesium (which has 12 electrons)   [1]

  e chlorine (which has 17 electrons)   [1]

  f calcium (which has 20 electrons).   [1]

  g **i** In what way is the group number of an element in the Periodic Table related to its arrangement of electrons?   [1]

    ii Which of the elements in **a** to **f** above is in Group 7?   [1]

  h **i** If an element is in the second period in the Periodic Table, how many electron shells (energy levels) will be occupied by at least one electron?   [1]

    ii Which element in **a** to **f** above is in the 4th period of the Periodic Table?   [1]

3 A group of students wanted to prepare a sample of crystals from the reaction between the insoluble solid, barium carbonate ($BaCO_3$) and dilute hydrochloric acid.

  a Give two ways in which the students could tell when all the acid had reacted.   [2]

  b Write down the general equation that describes the reaction between a carbonate and an acid.   [2]

  c Write the word equation for the reaction between barium carbonate and dilute hydrochloric acid.   [1]

  d Describe the method that the group should follow to obtain a good sample of crystals.   [5]

4 A student added a piece of magnesium ribbon to dilute hydrochloric acid.

  a List three ways by which she could tell that a chemical reaction was taking place. [3]

  b Write down the general equation that describes the reaction between a metal and an acid.   [2]

  c Write the word equation for the reaction between magnesium and dilute hydrochloric acid.   [2]

5 Write the word equation for each of the following reactions:

  a calcium + hydrochloric acid   [2]

  b zinc + sulfuric acid   [2]

  c magnesium carbonate + hydrochloric acid [3]

  d sodium carbonate + hydrochloric acid   [3]

  e calcium oxide + nitric acid   [2]

  f magnesium oxide + hydrochloric acid   [2]

  g sodium hydroxide + copper sulfate   [2]

  h potassium hydroxide + nitric acid   [2]

6 Zinc oxide is a white powder that is insoluble in water. It reacts with dilute hydrochloric acid.

  a What will you observe when the reaction has finished?   [1]

  b What type of substance do we call zinc oxide?   [1]

  c Write the word equation for the reaction. [2]

  d **i** Describe the sequence of steps needed to prepare crystals of the salt formed in the reaction.   [5]

    ii All sodium compounds are soluble in water. Explain why the method in part **i** would not work if you reacted sodium oxide with sulfuric acid.   [2]

7 Sanjay and Becky were preparing a sample of copper sulfate crystals. They started by reacting copper oxide with dilute sulfuric acid.

a What do we call the reaction between an acid and a base? *[1]*

b Put these steps in their method into the correct order.

A Pour 20 cm³ of sulfuric acid into a small beaker.

B Pour the solution into an evaporating dish. Heat it on a water bath.

C Warm the beaker gently on a tripod and gauze.

D Add more copper oxide, one spatula at a time and stir until no more will dissolve.

E Leave the solution at room temperature for a few days.

F Add a spatula of black copper oxide. Stir with a glass rod.

G Stop heating when you see a few small crystals appear around the edge of the solution.

H Filter off the excess copper oxide from the solution. *[3]*

c Write a word equation for the reaction between copper oxide and sulfuric acid. *[1]*

8 This question is about the elements in Group 7 of the Periodic Table. The elements are:

F – fluorine; Cl – chlorine; Br – bromine; I – iodine; At – astatine

a Draw an atom of a fluorine atom (atomic number 9), showing the arrangement of its electrons. *[1]*

b The Group 7 elements get less reactive going down the group. This group contains the most reactive of the non-metallic elements. Name this extremely reactive element. *[1]*

c Fluorine is a gas at 20 °C, as is chlorine. Bromine is a liquid and iodine is a solid at 20 °C. What state will astatine be in at 20 °C. Explain your reasoning. *[2]*

9 A compound whose old name is 'potash' (K_2CO_3) is reacted with nitric acid (HNO_3) to make potassium nitrate (KNO_3).

a i What is the chemical name for K_2CO_3? *[1]*

ii How many different types of atom are there in 'potash'? *[1]*

b Write a word equation for the reaction between 'potash' and nitric acid. *[2]*

10 Look at the atoms shown below. Their letters are NOT their chemical symbols.

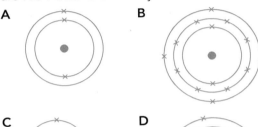

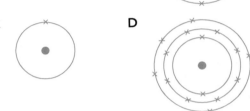

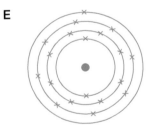

Which of A to E:

a is the lightest of all the elements? *[1]*

b belong in the same group of the Periodic Table? *[1]*

c has 13 protons in its nucleus? *[1]*

d is the second element in Group 5? *[1]*

e is in the 2nd period of the Periodic Table? *[1]*

f are atoms of non-metallic elements? *[2]*

g has an atomic number of 19? *[1]*

2 Reactivity and rates of reaction

Science *in context!* — Alloys

In this chapter we will start by looking at metals and their reactions. The metals will all be metallic elements. However, many important everyday metals are not elements but mixtures of metals called alloys.

An alloy is a mixture of a metal with the other elements, usually other metals. They are chosen by materials scientists to have the best properties for a particular job. To make the alloy, you simply melt the metals and mix them together. The metals do not react with each other. That's why we can describe the alloy as a mixture (and not a compound).

You can read about some examples below.

Common alloys

Name of alloy	Metals the alloy contains	Example of use
brass	copper and zinc	door handles
duralumin	aluminium, copper and magnesium	aeroplanes
solder	lead and tin	making electrical circuits

Some other alloys you use everyday are the coinage alloys. These alloys must be very hard-wearing. They also have to be malleable enough to be stamped with complex patterns.

Steels

Steels are alloys of iron – the most common metal used in construction. Normal steels have small amounts of carbon mixed with the iron. However, you can also get special alloy steels, such as stainless steel, which contains chromium and nickel mixed with the iron. A very hard steel is made by alloying iron with tungsten metal.

Why does alloying make a metal stronger?

We can use the particle model to explain how mixing metals in alloys makes them stronger.

With different sized atoms mixed in, the layers can't slip past each other so smoothly. It's a bit like little stones jamming a door!

Steel girders give many modern buildings their strength

Look at the diagram below:

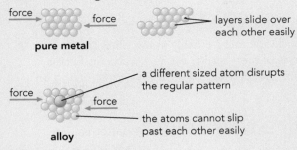

Memory metals

Some of the most recent 'smart materials' to be developed are called memory metals. These special alloys of nickel and titanium seem to 'remember' their original shape. They can switch back to this shape at a certain temperature. You might have seen memory metals advertised in the latest sunglasses.

Here are some additional uses of memory metals:
- joining hydraulic lines in fighter planes
- holding ligaments to bones
- supporting and keeping blood vessels open
- braces for teeth
- fire sprinklers
- thermostats on coffee pots.

Can you think of any other possible uses for memory metals?

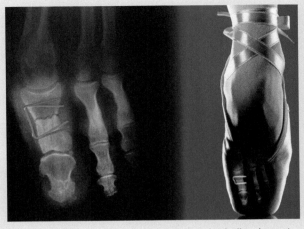

A memory metal is used to hold together this ballet dancer's fractured toe. The memory metal warms up to return to its original shape – which holds the bone in place with no gap.

In this chapter you will find out about the reactivity series of metals with oxygen, water and dilute acids, and about their displacement reactions. You will also learn about the effects of concentration, particle size, temperature and catalysts on the rate of a chemical reaction.

Key points

- We can judge the order of reactivity of metals by comparing their reactions with oxygen, water and dilute acid.
- The list of metals in order of reactivity is known as the Reactivity Series.
- A metal higher up the Reactivity Series can displace a less reactive metal from its compounds. For example, in solution:

   zinc + copper sulfate → zinc sulfate + copper

This is called a displacement reaction.
- Chemical reactions are speeded up by increasing:
   – surface area
   – concentration (or pressure if gases are reacting)
   – temperature.
- The collision theory explains why these factors affect the rate or speed of a reaction. When particles collide more often, reactions speed up.
- With higher temperatures, the collisions are also harder (more energetic). This means that more collisions in a given time have enough energy to produce a reaction.
- Some reactions are also speeded up by a catalyst. The catalyst itself is not chemically changed at the end of the reaction.

Learning outcomes

After this topic you should be able to:

- write word equations for the reactions of metals with oxygen
- describe patterns in reactivity of metals with oxygen.

In this chapter you will look more closely at the differences between the reactions of metals with oxygen, water and acids. You can then use these differences to set about constructing an order of reactivity.

Metals reacting with oxygen

Look at the photo below showing some freshly cut lithium. Compare it with the same piece of metal a few minutes later:

Freshly cut lithium

The same piece of lithium a few minutes later

Remember that lithium is one of the reactive Group 1 metals in the Periodic Table (see page 10). It reacts quickly with oxygen in the air.

Practical activity   Metals reacting in air and in oxygen

1 Your teacher will set up a dish containing a piece of copper, iron, magnesium and zinc metal. Notice that the metals are all sanded to expose their shiny surfaces to start with.

You can check the dish over the next few weeks to observe any changes if the metal reacts with oxygen in the air.

2 Watch your teacher heat samples of the same metals on a combustion spoon. As soon as the metal glows, it is plunged into a gas jar full of oxygen gas.

 Observe the magnesium reaction through tinted glass.

- Record your observations in a table.
- Explain your observations.
- Comment on the fairness of using your results to compare the reactivity of copper, iron, magnesium and zinc.

The air contains about 20% oxygen. Often a metal will form a dull coating of the metal oxide as it reacts with oxygen when exposed to air. The reaction of a metal with gases in the air causes its surface to **tarnish**.

For example, calcium is quite a reactive metal:

calcium + oxygen → calcium oxide

We find that metals, if they react with oxygen, do so at different rates. Some, such as lithium and potassium, react quickly. Others, such as copper, react slowly.

When copper foil is heated strongly in a Bunsen flame it forms a coating of black copper(II) oxide on its surface:

$$\text{copper} + \text{oxygen} \xrightarrow{\text{heat}} \text{copper(II) oxide}$$

Gold is so unreactive that it doesn't react in air and stays shiny, even when heated.

Gold is such an unreactive metal that it will not react with oxygen in the air even when heated to its melting point

Expert tips

When the surface of a metal turns dull in the air we say that the metal 'tarnishes'. Tarnishing is an oxidation reaction.

Other gases in the air, besides oxygen, can also be involved in the tarnishing of metals. These gases include carbon dioxide, water vapour and nitrogen. Pollutants such as sulfur dioxide, may also cause tarnishing.

Key terms

- **tarnish**

Summary questions

1 Copy and complete:

   When a metal reacts with gases in the _____ its surface loses its _____ appearance. We say that the metal's surface gets _____ . Many metals react slowly with _____ in the air to form oxides.

2 Copy and complete these word equations:

   a) magnesium + oxygen → _____

   b) _____ + _____ → aluminium oxide.

3 Put the following five metals in order of reactivity judging from their reactions with oxygen, with the most reactive first:

   iron   gold   copper   magnesium   lithium

Most metals don't react vigorously with water. Just think of the cooking pans in your kitchen at home! However, as you found with tarnishing in air, there are differences in the reactivity of different metals.

Practical activity	Comparing magnesium and copper in cold water

Clean the surface of a strip of magnesium ribbon with emery paper.

Wind the magnesium ribbon into a coil and place it in the apparatus shown below.

Repeat this with a strip of copper foil.

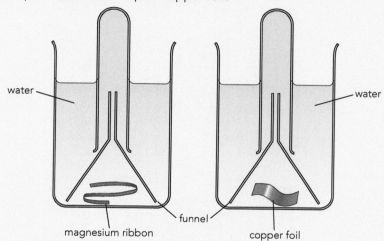

Leave the apparatus for a week.

- What do you observe after a week? Which metal shows no sign of a reaction?

Test any gas collected with a lighted splint.

- What happens? Which gas is produced?
- Where did this gas come from?
- Which of the two metals is more reactive with water – magnesium or copper?

Using water to compare the reactivity of metals is difficult because many of them only react very slowly with cold water. This makes any differences in the rates of reaction hard to observe or measure.

Some fairly reactive metals will not react with cold water but will react with water when it is in the gaseous state. If we pass steam over the metals, a reaction does take place. These metals are magnesium, iron and zinc. The metal oxide is formed and hydrogen gas is produced. (See page 38.)

For example:

magnesium + steam → magnesium oxide + hydrogen

However, with the most reactive metals we can use the reaction with cold water to judge relative reactivity. We can see the different rates at which hydrogen gas is given off quite easily. These reactive metals produce the metal hydroxide in solution and give off hydrogen gas. The general equation is:

reactive metal + water → metal hydroxide + hydrogen

Putting the reactive metals in order

Four of the most reactive metals we can use in school science are lithium, sodium, calcium and potassium. Your task in the next experiment is to use your observations to place these reactive metals into their order of reactivity.

Practical activity   Reactive metals and water

Watch your teacher demonstrate the reactions of lithium, sodium, calcium and potassium with water.

 Observe from behind a safety screen and wear eye protection.

- Record, in detail, your observations of each reaction.
Include the name of the gas given off and the pH of the solutions formed.
- In what ways were the reactions similar?
- In what ways were the reactions different?
- What safety precautions did your teacher take when using these metals and demonstrating their reactions?
- What is the hazard warning sign on each jar of the metals? Why?
- Why are the lithium, sodium and potassium stored under oil in their jars?
- Put the metals in order of reactivity with water, listing the most reactive one first.

The solutions formed in the reactions of reactive metals with cold water are alkaline because of the soluble hydroxides made:

lithium + water → lithium hydroxide + hydrogen

Caesium (Cs) is an extremely reactive metal. It reacts violently with water, smashing the glass trough.

Summary questions

1 Copy and complete:

The most reactive metals react _____ with cold water. They give off _____ gas and form _____ solutions of the metal _____. _____ was the most reactive metal we tested with water.

2 a) Find out the name of the metal in Group 1 of the Periodic Table whose symbol is Rb.
 b) Predict what you would observe when Rb reacts with cold water.
 c) Write a word equation for this reaction.

Metals plus acid – how fast?

After this topic you should be able to:

- place metals in order of their reactivity using their reactions with dilute acid, water and oxygen
- plan and carry out safe tests comparing the reactions of different metals with dilute acids
- evaluate your investigations.

Magnesium reacts with dilute acids, giving off hydrogen gas and forming a solution of a magnesium salt. The name of the salt depends on the acid used in the reaction, e.g. with dilute sulfuric acid, the salt formed in solution is magnesium sulfate.

Expert tips

Aluminium metal appears to be less reactive than its position in the Reactivity Series would suggest. This is because its surface is covered in a tough, non-porous layer of aluminium oxide which protects the aluminium metal beneath it.

What's the order?

You have now seen how a range of metals react (or don't react) with oxygen and water. We have used our observations of these reactions to place some metals into an order of reactivity. But with some metals the reactions are very slow. This makes it difficult to put them in order of reactivity. With these metals we can look at their reactions with dilute acid to judge their positions in an order.

Practical activity   Judging reactivity using dilute acids

You are given coarse-mesh filings of the metals copper, zinc, iron and magnesium. Your task is to put them into an order of reactivity using their reactions with either:

- dilute hydrochloric acid,
- dilute sulfuric acid, or
- dilute ethanoic acid.

Plan a test to put the metals in order of reactivity based on your observations. Your plan should include the quantities of reactants you intend to use. (You will have access to a balance and measuring cylinders.)

In this investigation you should plan to collect data to test out your order of reactivity. Measurements should play a big part in the results you record. This is called **quantitative data** (involving numbers) as opposed to the **qualitative data** that are based on observations.

The 'type of metal' will be your **independent variable**.

Think about these questions to help your planning:

- What could you measure as each reaction proceeds? This will be your **dependent variable**.
- Which variables will you keep constant (**control variables**) to make your tests as fair as possible?
- What steps will you take to make sure your tests are safe?
- In what way will you record your results?
- In what way will you display your results?

⚠ **Let your teacher check your plan before you start your practical work.**

Following your practical tests:

- Put the metals tested into an order of reactivity – with the most reactive first.
- In what way, if any, do the different acids affect the reactions?
- Evaluate your investigation. Think about the quality of the data collected and explain any ways you could improve your method.

The Reactivity Series of metals

The order of reactivity of metals is called the **Reactivity Series**. The table below shows the reactions of metals with dilute acid. Potassium, sodium and lithium are all above magnesium in the Reactivity Series but are not safe to mix with dilute acid, exploding on contact.

Key terms

- qualitative data
- quantitative data
- Reactivity Series

Metal (in order of reactivity)	Reaction with dilute acid
magnesium	bubbles of hydrogen gas are given off quickly, and a salt is formed in solution
aluminium	gives off hydrogen gas, and a salt is formed in solution (but only if the surface is cleaned first to remove its oxide layer)
zinc	gives off hydrogen gas, and a salt is formed in solution
iron	gives off hydrogen gas, and a salt is produced in solution
tin	only reacts slowly when the acid is warm
copper	no reaction
silver	no reaction

Here is a summary showing how all the metals we have looked at in this chapter react with oxygen and water.

Reactivity Series	Reaction with oxygen	Reaction with water (or steam)
potassium, sodium, lithium	rapidly tarnish in air when heated they burn with a bright flame to form white oxide powders	react vigorously in cold water, giving off hydrogen gas and forming a strongly alkaline solution of the metal hydroxide
calcium		reacts steadily in cold water, giving off hydrogen gas and forming a slightly less alkaline solution of calcium hydroxide
magnesium		reacts very slowly with cold water, but vigorously in steam, giving off hydrogen gas
aluminium	tarnishes in air, forming a thin coating of non-porous aluminium oxide (which protects its surface)	if the oxide coating is removed, it will react slightly with water, giving off hydrogen; aluminium oxide will form on its surface preventing further reaction
zinc	zinc oxide forms on its surface slowly but more rapidly if heated	hydrogen released from steam zinc oxide formed
iron	forms an oxide if heated; rusts in the cold, if water is also present	reacts with steam to produce hydrogen iron oxide formed
tin	forms an oxide when heated	no reaction with cold water and only a slight reaction with steam
copper	forms black copper oxide but must be strongly heated	no reaction
silver	no reaction	no reaction

Summary questions

1 Copy and complete: 'Metals can be listed in ___ of reactivity using their reactions with ___, dilute acid and ___. This is called the ___ Series.'

2 Explain why aluminium appears to be in a false position in the Reactivity Series.

2.4 Displacement reactions

Learning outcomes

After this topic you should be able to:

- explain displacement reactions
- predict displacement reactions using the Reactivity Series
- write word equations for displacement reactions.

Displacement from solution

So far, we have put the metals in order of reactivity on the basis of their reactions with water, dilute acids and oxygen. Now we can use the resulting Reactivity Series to predict what happens when we put the metals 'into competition' with each other.

Practical activity Metals in competition

Set up the boiling tubes as shown:

Leave the metals and solutions to react for about 15 minutes.

- Record your observations.
- Think about the reactants in each tube and what they could form. (Remember that chemical reactions involve a rearrangement of the atoms in the reactants to form the products.)
- Try to explain in your own words what might be happening in each reaction.

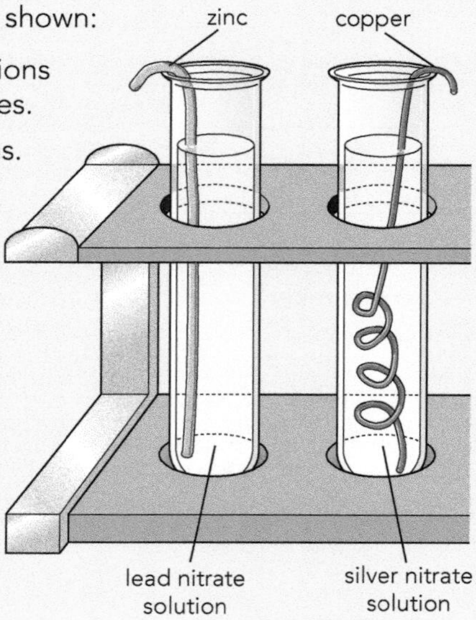

Wash your hands after using lead nitrate.

Let's look at an example using a 'model' involving competition between metals:

If we have silver nitrate solution and copper, we have set up a competition between silver and copper. Both copper and silver could form a compound, but it is silver that starts off with the nitrate as a compound in solution.

However, copper is more reactive than silver. It appears above silver in the Reactivity Series. We can think of it as 'stronger' than silver. Therefore it 'takes the nitrate' for itself, going into solution to join it. The silver is 'kicked out' of solution and left as the solid metallic element, silver itself.

Remember that this is just a model to help us understand what is happening.

The word equation is:

copper + silver nitrate → copper nitrate + silver

We call this a **displacement reaction**.

If we added silver to copper nitrate, there would be no reaction.

Silver is not reactive enough to displace copper from its solution.

silver + copper nitrate → no reaction

Practical activity	Predicting displacement from solution

You are given small samples of the following metal powders:
- magnesium
- copper
- iron
- zinc

as well as access to solutions of:
- magnesium sulfate
- copper sulfate
- iron(II) sulfate
- zinc sulfate.

- Draw a table that is large enough to record your observations of the different combinations of each metal with each sulfate solution. The table should also have space for your predictions of whether you expect to see a reaction (YES) or not (NO).

You can observe the reactions on a spotting tile. Use the end of a wooden splint to add a small amount of the metal powder to the well containing the sulfate solution you are testing.

 Wear eye protection.

- Were there signs of a reaction in all the combinations that you predicted as 'YES' in your table?

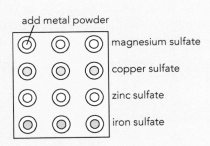

- Select four reactions from your table and write word equations for these displacement reactions.

We find that a more reactive metal will displace a less reactive metal from a solution of one of its salts.

For example:

zinc + copper sulfate → zinc sulfate + copper

Expert tips

When explaining the displacement reaction between zinc and copper sulfate, an ideal answer would be:

'Zinc is **more reactive** than copper, therefore it can **displace** the copper from copper sulfate solution, forming copper metal and leaving zinc sulfate in solution.'

Do NOT say that zinc is 'stronger' than copper! This is just a model to help you understand why displacement reactions take place.

Key terms

- **displacement reaction**

Summary questions

1 Copy and complete:
  A _____ reactive metal will _____ a _____ reactive metal from its compounds.
  We call this a _____ reaction.

2 Write a word equation for the reaction between iron and copper sulfate.

3 Explain what would happen if you added iron to zinc sulfate solution.

The reactivity of a metal is related to:

- its occurrence (where we find the metal in nature)
- its uses
- its ease of extraction
- when it was discovered.

We have seen that many metallic elements react with oxygen, water and dilute acids. Oxygen, water and dilute acids are present in our environment. So most metals are found in nature as compounds; combined with elements of non-metals. Only a metal very low down in the Reactivity Series can exist as the element itself. Gold is an example. We can find gold in rocks or in rivers that have worn away the rock. To get the gold, the small amount of metal has to be separated from tonnes of rock.

Gold can be found as the element in nature

 IGCSE Link...
You will find out more about the extraction of metals in the IGCSE Chemistry course.

Most metals we use occur as compounds found in **ores**. Ores are rocks from which it is economically worthwhile to extract the metal. Remember that metals are bonded to non-metals in their compounds. So the metals must be extracted from compounds in their ores by using chemical reactions. We find generally that:

the more reactive a metal is, the more difficult it is to extract from its ore.

Key terms

- **ore**

The highly reactive metals need to be extracted by electrolysis of one of their molten compounds found in an ore.
Once we have samples of the metal, the reactivity of the metal will be one of the factors that determine its uses.

Practical activity   Finding links with the Reactivity Series

Working as a group, collect together some examples of:
- uses of metals
- the occurrence of metals in nature
- the way we extract metals from their ores
- the dates of discovery.

Relate these points to the metal's position in the Reactivity Series.

You can use a range of source materials such as books, DVDs, CD ROMs, posters, leaflets and the internet.

- Write an account to summarise your findings in each area. Make sure that you structure your work logically and use paragraphs and illustrations to develop each idea.

Summary of the reactivity of metals

Here is a summary of the reactions we used to determine the Reactivity Series of metals:

Order of reactivity	Reaction when heated in air	Reaction with water	Reaction with dilute acid
potassium	burn brightly, forming metal oxide	fizz in cold water, giving off hydrogen, leaving an alkaline solution of metal hydroxide	explode
sodium			
lithium			
calcium			fizz, giving off hydrogen and forming a salt
magnesium		react with steam, giving off hydrogen and forming the metal oxide	
aluminium			
zinc			
iron			
tin	oxide layer forms without burning	slight reaction with steam	react slowly with warm acid
lead			
copper		no reaction, even with steam	no reaction
silver	no reaction		
gold			

Order of reactivity of metals

Summary questions

1 Copy and complete:
   Only metals which are _____ can be found in nature as the _____ itself.
   Other metals have to be _____ from their _____ by chemical reactions, including by electrolysis.

2 Gold was one of the earliest metals extracted from the ground. Explain why.

3 Explain why platinum (Pt) is used to make jewellery and magnesium (Mg) is used in fireworks.

Learning outcomes

After this topic you should be able to:

- describe the effect of increasing the surface area of a solid on the rate of reaction
- explain why surface area affects the rate of a reaction using the collision theory
- assess hazards in an experiment.

Have you ever tried to burn wood on a fire? Which would burn more quickly – a block of wood or the same block of wood chopped into small pieces? We find that small pieces of solids, especially powders, react faster than large pieces of solids.

Why is it best to start a fire with small pieces of wood?

Surface area is a measure of how much surface of a solid is exposed. Small pieces of a solid have a larger surface area than the same mass of the solid in large lumps.

Imagine deep frying two pans of chips. One pan has potato cut into small, thin chips. The other pan has bigger, thicker chips. So for the same mass of potato, small chips have a larger surface area than big chips.

Which chips will be cooked first? Which chips have the larger surface area?

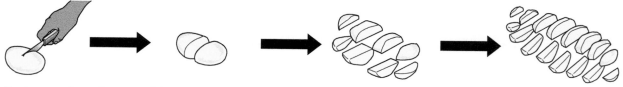

You increase the surface area of the potato each time you cut it smaller

Let's look at the effect of increasing surface area on the rate of a chemical reaction in the Practical activity on page 35.

Practical activity   The effect of increasing surface area on rate of reaction

Compare what happens when you heat:
- an iron nail
- iron wool
- iron filings.

Hold the iron nail and iron wool in tongs as you heat them.

Gently sprinkle a few iron filings into a Bunsen flame from the end of a spatula.

 Eye protection must be worn.

- Put the three types of iron in order of increasing surface area.
- What effect does increasing the surface area of the iron have on the rate of its reaction?
- What were the risks to yourself and others in this experiment? What did you do to control these to make sure your experiment was safe?

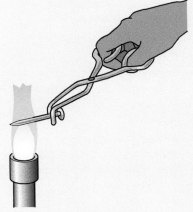

Write a word equation for the reaction of iron with oxygen in the air

The collision theory

As you know, all substances are made up of particles. The particles might be atoms, molecules or charged particles called ions. Before we can get a chemical reaction, particles must crash together. They must collide with enough energy to cause a reaction. This is called the **collision theory**.

Think about the rate of a reaction. What do you think happens to the number of collisions in a certain time between particles if you speed up a reaction?

Think about burning iron filings in the previous experiment. Powders have a very large surface area. There are lots of iron atoms exposed at its many surfaces. The oxygen molecules in the air can attack any of these iron atoms. With iron filings, there are lots of collisions in a given time. Therefore the chemical reaction is very fast.

Compare this with heating the iron nail. The iron nail has a small surface area. It only reacts slowly.

Expert tips

When you explain increasing 'rates of reaction' make sure you write about increasing the number of collisions between reacting particles IN A GIVEN TIME.

Just saying 'There are more collisions' is not enough, as these collisions could take place over a long period of time!

Key terms

- **collision theory**

 IGCSE Link...
The work you do on rates of reaction will be revisited when you do IGCSE Chemistry.

Summary questions

1 Put the following in order of increasing surface area (lowest surface area first). Assume you have the same mass of magnesium in each case:

   magnesium ribbon   **magnesium powder**   **magnesium bar**

2 Explain why magnesium powder burns in a Bunsen flame more quickly than magnesium ribbon.

3 We can burn lumps of coal safely in domestic fires. So why is the coal dust in the air inside a coal mine dangerous?

Do you like your orange squash strong or weak? Look at the three glasses opposite:

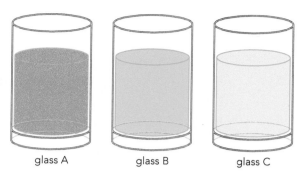

glass A   glass B   glass C

The glasses contain watered-down orange squash. The glasses all contain the same volume of solution. Glass A contains more orange squash and less water than the other glasses. We say that the orange squash solution is more concentrated in glass A than in glass B or C.

The orange squash in glass A is more concentrated because it has more particles of orange squash in the same volume of solution:

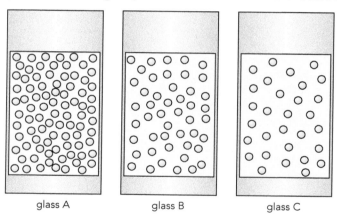

glass A   glass B   glass C

The acids we use in science are usually watered down (diluted). Different proportions of acid to water make solutions of different **concentrations**.

In the following experiment you can find out how the concentration of acid affects the rate of reaction with magnesium.

Practical activity   The effect of concentration on rate of reaction

You know what happens when we add excess dilute acid to magnesium ribbon. Hydrogen gas is given off and the magnesium gets smaller and smaller. Eventually all the magnesium has reacted and we are left with a solution of a magnesium salt.

In this experiment you will **vary the concentration** of hydrochloric acid each time you do a test.

- What other variables must be kept constant to make this a fair test?

You can measure how quickly the reaction takes place by timing how long it takes for a 3 cm long piece of magnesium ribbon to disappear. Start timing as soon as the magnesium ribbon is dropped into a small beaker containing the acid.

You can dilute the most concentrated acid solution provided with water in the proportions shown in the table.

- Copy and complete the table with your results.

Test number	Acid / cm³	Water / cm³	Time for reaction to finish / s
1	10	40	
2	20	30	
3	30	20	
4	40	10	
5	50	0	

⚠ **Eye protection must be worn.**

- Which test used the most concentrated of the hydrochloric acid solutions?
- In which test was the reaction fastest?
- In what way could you set up a test to act as a 'control' experiment?
- What pattern can you see in your results?

Experiments such as the previous one show us that:

As concentration increases, the rate of reaction increases.

This happens because in a concentrated solution there are more dissolved particles in the same volume of solution. Remember that the particles in any liquid are in constant random motion. So when they are more crowded together they will collide with other reactant particles more frequently. Therefore the reaction happens more quickly.

Look at the diagram below:

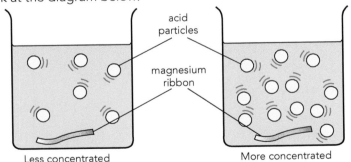

Less concentrated | More concentrated

More acid particles will collide with the magnesium ribbon in a given time in the beaker with the more concentrated acid. Therefore the reaction will be faster.

Summary questions

1 Copy and complete:

As we increase the concentration of a _____ , the rate of reaction _____ . This is because there are more dissolved _____ in the same _____ of solution so there will be more frequent _____ .

2 What do we mean when we say that one solution is more concentrated than another?

3 In the reaction between magnesium ribbon and dilute hydrochloric acid in the experiment above, why did the reaction stop in each test?

We can keep food longer at low temperatures

Why is food stored at low temperatures in fridges or freezers? Some of the substances in food react with oxygen in the air. Oils and fats in many foods turn 'rancid' when left in air. They react and turn into acids. The **low temperature** in a fridge **slows down** the reactions that make food go bad.

You have seen the effect that temperature can have on a reaction when studying the reactions of metals with water. Magnesium ribbon reacts very, very slowly with cold water. However, if magnesium is heated and water in the form of steam is passed over it, the reaction is vigorous.

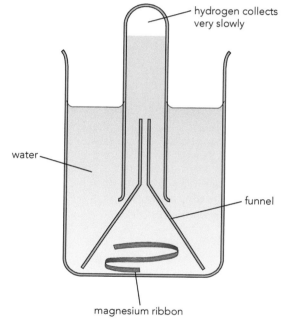

Magnesium reacting very slowly with cold water

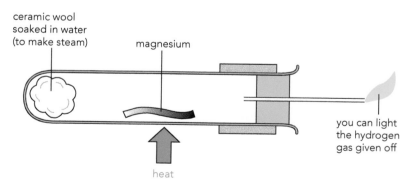

Magnesium reacting very quickly with steam

You can see the effect of temperature on the reaction between magnesium ribbon and dilute sulfuric acid in the next experiment.

Practical activity   The effect of temperature on rate of reaction

You know that when we add excess dilute acid to magnesium ribbon, hydrogen gas is given off and the magnesium gets used up. In this experiment you will use dilute sulfuric acid and *vary its temperature* each time you do a test.

 Eye protection must be worn.

- What variables must be kept constant?

Place 20 cm³ of dilute sulfuric acid in a boiling tube and take its temperature.

Drop in a 1 cm length of magnesium ribbon and time how long it takes to get used up.

Use a water bath to warm another 20 cm³ of acid up to 30 °C. Time how long it takes for the magnesium to get used up. Repeat the same test at 40 °C.

- Record your results in a table.
- What pattern do you see in your results?

Experiments such as the previous one show us that temperature has a large effect on the rate of reaction:

As temperature increases, the rate of reaction increases.

This happens because as a solution is warmed up, its particles gain energy and move around more quickly. Therefore they will collide with other reactant particles more frequently. This will make the reaction happen faster.

Not only that, the particles collide with more energy at a higher temperature. The reacting particles need a certain amount of energy before a collision results in a reaction taking place. So at higher temperatures there are a higher proportion of particles with enough energy to react.

Look at the diagram below.

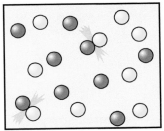
Reaction at 30°C

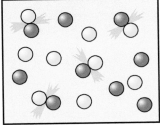
Reaction at 40°C

Expert tips

In many reactions, as the temperature is increased by just 10 °C the rate of reaction doubles.

Summary questions

1 What happens to the particles in a solution when they are heated up?

2 Write a word equation for the reaction between magnesium and dilute sulfuric acid.

3 Explain why temperature affects the rate of a chemical reaction.

Chemists have discovered that some substances can make certain chemical reactions happen more quickly. These substances are not changed chemically themselves at the end of the reaction. We call these substances **catalysts**.

In the following experiment you can see what effect a catalyst has on the rate of a reaction.

Practical activity   Breaking down hydrogen peroxide

 Eye protection must be worn.

Pour some hydrogen peroxide solution into a boiling tube.
- What do you see around the inside of the tube?
- Are the bubbles forming quickly?

The solution gives off oxygen gas as it breaks down.

Try testing for oxygen gas with a glowing splint.
- What happens?

Now add a little manganese(IV) oxide, and test for oxygen again.
- What happens as soon as you add the manganese(IV) oxide?
- Did your glowing splint re-light this time?

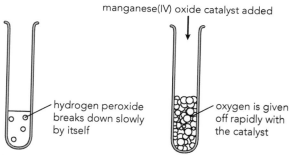

manganese(IV) oxide catalyst added

hydrogen peroxide breaks down slowly by itself

oxygen is given off rapidly with the catalyst

The effect of manganese(IV) catalyst on the breakdown of hydrogen peroxide

The manganese(IV) oxide is a catalyst for this reaction:

$$\text{hydrogen peroxide} \xrightarrow{\text{manganese(IV) oxide}} \text{water} + \text{oxygen}$$

We say that the manganese(IV) oxide **catalyses** the reaction. Notice that the catalyst does not actually appear in the word equation. We can write it above the arrow to show that it is there during the reaction.

Investigating catalysts

Catalysts work because they can lower the minimum amount of energy that colliding particles need in order to react. So there are more particles with sufficient energy to react when a catalyst is used.

To find out which substances will act as a catalyst for a particular reaction, chemists have to try them out and see what happens. In the following investigation you can find the best catalyst for the breakdown of hydrogen peroxide.

Practical activity   Which is the best catalyst to break down hydrogen peroxide?

Your task is to find out which **metal oxides** will catalyse the breakdown of hydrogen peroxide.

- What do you predict will happen?
- What will you do to make it a fair and safe test?
- In what way will you judge how quickly the hydrogen peroxide breaks down? Can you take measurements rather than just relying on observations?

 Check your plan with your teacher before you start any practical work.

Catalysts in industry

Chemists have to find catalysts for reactions that take place in industry. The quicker a product can be made the greater the profit a company can make.

Modern cars have catalytic converters fitted in the exhaust systems to reduce pollution. The catalyst is in the form of a honeycomb in order to increase its surface area. This makes the catalyst more effective.

Key terms

- **catalysts**

Practical activity   Which catalysts are used in industry?

Carry out some research using secondary sources to find some industrial processes that use catalysts.

- Write a word equation, including the catalyst, for each reaction you find.
- Make a list of the benefits of using catalysts in industry.

Summary questions

1 Define the word catalyst.

2 Explain the way in which a catalyst can increase the rate of a reaction.

3 a) In a word equation, why is the catalyst for a reaction not shown as a reactant or product?
    b) In what ways can we show the catalyst in a word equation? Give an example.

1 Here are some reactions of four metals:

Metal	With cold water	With dilute sulfuric acid
iron	no immediate reaction	starts fizzing, giving off a gas
silver	no reaction	no reaction
sodium	floats, then melts as it skims across the surface of the water, giving off a gas	too dangerous to attempt
nickel	no reaction	a few bubbles of gas are given off if the acid is warmed

a List the four metals in their order of reactivity (most reactive first). [1]

b **i** Name another metal that reacts in a similar way to sodium. [1]

ii Which gas is given off in the reaction between sodium and water? [1]

iii What will be the pH of the solution that remains after the reaction between sodium and water?

    1   3   7   14 [1]

iv Why should the reaction between sodium and dilute sulfuric acid not be tried in a school laboratory? [2]

c **i** Name the gas given off when iron reacts with dilute sulfuric acid. [1]

ii What is the other product formed in this reaction. [1]

d Look at the test tubes below:

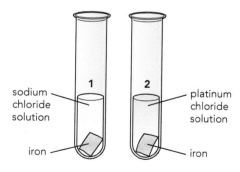

sodium chloride solution — 1

platinum chloride solution — 2

iron

iron

In test tube 2 the iron was slowly covered in a light grey deposit.

    i Name the light grey deposit in test tube 2. [1]

ii Explain why the light grey deposit formed. [1]

iii What do we call the type of reaction observed in test tube 2? [1]

iv Write a word equation for the reaction in test tube 2. [1]

v Why did no reaction take place in test tube 1? [1]

2 This question is about four metals – labelled A, B, C and D (these are not their chemical symbols). The metals and solutions of their sulfates were mixed in a spotting tile:

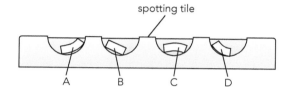

spotting tile

A   B   C   D

The table below shows the results:

✓ shows that a reaction took place

✗ shows that no reaction took place

	A	B	C	D
A sulfate	✗	✓	✓	i)
B sulfate	✗	✗	✓	✓
C sulfate	✗	ii)	✗	iii)
D sulfate	✗	✗	✓	✗

a Use the table to list the metals A, B, C and D in order of reactivity (most reactive first). [1]

b Use your answer to a) to decide whether a ✓ or a ✗ goes in the boxes i), ii) and iii) in the table. [2]

c Copper metal reacts with silver nitrate solution.

    i Write the word equation for this reaction. [1]

ii Platinum does not react when placed in a solution of silver nitrate.

List the three metals – platinum, copper and silver – in order of reactivity. Start with the most reactive. [1]

d Copper is used to make household water pipes and iron is used to make boilers.
Explain which of these two metals will corrode first. *[1]*

3 This table shows when some metals were discovered :

Metal	Known since:
potassium	1807
sodium	1807
zinc	before 1500 in India and China
copper	ancient civilisations
gold	ancient civilisations

a What pattern can you see between a metal's place in the Reactivity Series and its discovery? *[1]*

b Can you explain this pattern? *[2]*

4 A group of students are studying the effect of surface area on rate of reaction.
They use marble chips (calcium carbonate) reacting with dilute hydrochloric acid.

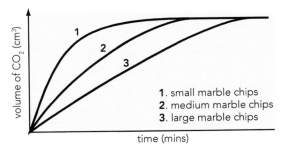

1. small marble chips
2. medium marble chips
3. large marble chips

a What did the students investigating the reaction have to do to make it a *fair test*? *[2]*

b Which size of marble chips has the largest surface area (given the same mass of each)? *[1]*

c Which marble chips reacted fastest? Explain your answer. *[2]*

d The students doing the experiments also tried reacting the same mass of *powdered* calcium carbonate with the acid. What would their results look like on the graph above? Copy the graph and clearly label the line for the calcium carbonate powder. *[2]*

5 a What is a **catalyst**? *[2]*

b Hydrogen peroxide decomposes to form water and oxygen gas. The reaction is catalysed by some metal oxides. You are given oxides of copper, manganese and nickel.
Describe a method you could use to test which is the best catalyst. *[5]*

c State the technique that could be used to separate the metal oxides from the reaction mixture. *[1]*

6 In countries where the climate can be cold, some houses are heated using wood-burners. These burn wooden logs:

When lighting a wood-burner, the manufacturers recommend lighting paper beneath a 'pyramid' of kindling (thin pieces of wood). Then larger logs can be added to the fire gradually.

kindling

paper

a Explain why kindling is used to start the fire. Use the collision theory in your answer. *[3]*

b Explain why the kindling is arranged in a 'pyramid' above the burning paper. *[2]*

7 Using the collision theory explain why the reaction of zinc granules with dilute hydrochloric acid is faster at 30 °C than at 20 °C. *[4]*

Science *in context!*

Energy from chemicals

Heating things up

We use the heat released in chemical reactions whenever we burn a fuel.

The heat energy released from burning charcoal is used to cook the food on this barbecue

However, we can also use the energy released in chemical reactions in different ways.

There are hand-warmers which use a super-saturated solution in a bag. When the bag is squeezed two pieces of metal rub together and tiny grains of metal scrape off. Crystals start forming in the solution, started by the tiny pieces of metal, and lots of heat energy is released.

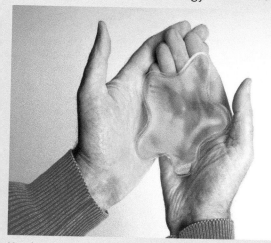

Hand-warmers are useful when watching sports in cold weather

Self-heating cans often use the energy released from a chemical reaction to generate the heat to warm up their contents. This has been used commercially to heat coffee in a can wherever you want to drink it.

A self-heating can of coffee. Pressing a button in the bottom of the can releases water onto calcium oxide. The reaction between water and calcium oxide releases lots of heat energy.

Cooling things down

Sports injuries can be treated with cold packs to reduce swelling and numb pain. These packs can contain ammonium nitrate and water.

The two are kept separate until the pack is squeezed. Then the ammonium nitrate dissolves in water, taking in energy. The instant cold packs work for about 20 minutes.

The cold packs are ideal for use on sports injuries in locations where access to ice is not possible.

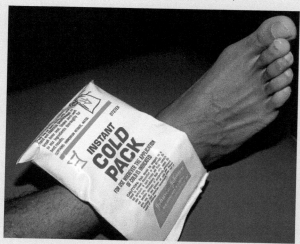

An ankle injury is treated with a cold pack to reduce swelling

Putting out fires

People sometimes heat oils or fats above their flash points. The flash point of a substance is the temperature at which it ignites. This happens in chip-pan fires. If water is thrown on to burning oil or fat, it makes the fire worse. The water turns to steam, and sends burning oil flying into the air. Look at the photo below:

Why does the damp tea-towel put out the fire? Why must you leave the pan covered by the tea-towel for some time after the fire?

In this chapter you will find out more about the energy changes that take place in chemical reactions, and the energy released when fuels burn.

Fireworks

The stunning effects in a fireworks display involve burning plenty of gunpowder. The Chinese were the first to use gunpowder. They used firecrackers in religious festivals.

To get the fierce reactions in fireworks you need a rapid supply of oxygen, which is provided by a chemical oxidising agent. The bright colours are produced by metal compounds in the firework mixture. You might remember the colours you got in previous work you did using flame tests for metals.

Metal compounds produce the bright colours in fireworks

Key points

- Exothermic reactions give out heat energy. The temperature rises.
- Endothermic reactions take in heat energy. The temperature falls.
- Three things are needed for a fire – fuel, heat, and oxygen.
- If one part of this 'fire triangle' is removed, the fire goes out.
- The reaction of a fuel with oxygen is called combustion.
- When a hydrocarbon fuel burns in a good supply of oxygen, it forms carbon dioxide and water.
- If a hydrocarbon fuel burns in a limited supply of oxygen, it can make carbon monoxide gas. This is called incomplete combustion.
- We can compare the energy content of different fuels in investigations in which we heat a fixed volume of water with each burning fuel.
- Respiration is an exothermic process whereas photosynthesis is an endothermic process.

Chemical energy into heat energy

In Chapter 2 we looked at the Reactivity Series of metals. On page 30 we saw how a more reactive metal can displace a less reactive metal. We used metals and solutions of their salts to investigate displacement reactions.

These displacement reactions, like combustion reactions, are **exothermic**. They release stored chemical energy as heat energy. So in an exothermic reaction the temperature of the surroundings increases as the reaction takes place.

For example:

copper sulfate + magnesium → magnesium sulfate + zinc

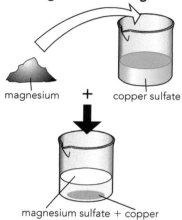

magnesium   **+**   copper sulfate

magnesium sulfate + copper

The displacement reaction between copper sulfate and magnesium releases energy to the surroundings

Absorbing energy

Not all reactions are exothermic, giving out energy. Some reactions take in energy from their surroundings and so the temperature falls. These are called **endothermic** reactions.

Practical activity   Exothermic or endothermic?

Use the apparatus as shown:

⚠ **Eye protection must be worn.**

stir with a thermometer

poly(styrene) beaker

Record the maximum or minimum temperature in each reaction using a sensitive thermometer

Mix each pair of substances in turn and record your results in a table like the one shown below:

Reaction	Temp. before mixing / °C	Temp. after mixing / °C
magnesium + dilute hydrochloric acid		
sodium hydrogencarbonate + dilute hydrochloric acid		
sodium hydroxide solution + dilute hydrochloric acid		
potassium hydrogencarbonate + dilute hydrochloric acid		

- Look at your results and classify each reaction as exothermic or endothermic.

You can now plan and carry out an investigation into the heat energy changes in displacement reactions.

Practical activity   Investigating heat energy from displacement reactions

You can investigate the energy given out during displacement reactions in this enquiry.

Use the same metals and solutions as the experiment 'Predicting displacement from solution' on page 31.

- Plan an investigation to find out which reactions give out most energy.
- Which measuring instrument can you use to measure the release of energy?
- In what ways will you make your tests as fair as possible?
- In what ways can you ensure your results are as valid as possible?

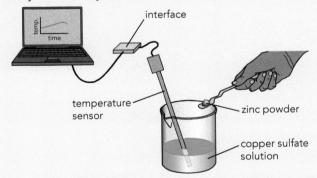

A thermometer or a temperature sensor can be used to measure the maximum temperature change in an exothermic reaction

 Do not start any experiments before your plan has been checked by your teacher.

Key terms

- **endothermic**
- **exothermic**

Summary questions

1 Explain what we mean by an exothermic reaction and an endothermic reaction.

2 Give an example of a reaction that is exothermic and one that is endothermic.

3 Two solutions, both at 19 °C, are mixed and a new temperature of 17.5 °C is recorded. Explain these observations.

3.2 Combustion

Learning outcomes

After this topic you should be able to:

- write a word equation for a combustion reaction
- form conclusions using your scientific knowledge and understanding.

The air around us is a mixture of gases. The two main gases in the air are nitrogen and oxygen. As you have seen from earlier work on burning, it is the oxygen that is important in burning.

Burning is a chemical reaction. The substance that burns reacts with oxygen. In this chemical reaction new substances called oxides are formed. Energy is also released during the exothermic reaction.

Burning is also known as **combustion**. A fuel releases energy as it reacts with oxygen in a combustion reaction.

We can investigate what is formed when a fuel burns in the next experiment.

Practical activity   Burning wax

Watch your teacher carry out the experiment below to test the gases given off when wax burns:

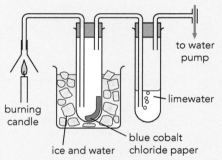

You already know the limewater test for carbon dioxide. You also need to know the test for water:

Test for water: blue cobalt chloride paper turns pink (or white anhydrous copper sulfate turns blue) in the presence of water.

- What do you see happen to the limewater in the experiment? What does this show?
- What happens to the blue cobalt chloride paper? What does this show?
- What are the products formed when wax burns in air?
- Why is the first tube surrounded by ice?
- Why do we use an upturned funnel and a water pump in this experiment?

Explaining combustion

- When a fuel containing carbon burns, the carbon turns to carbon dioxide (as long as there is plenty of oxygen around to react with).
- As well as carbon, most fuels also contain hydrogen. This hydrogen in the fuel molecules reacts with oxygen in the air to form water (which is hydrogen oxide).

Wax is made up of molecules made up of carbon and hydrogen. So when wax burns it produces carbon dioxide and water in its combustion reaction.

We can show this reaction in a word equation:

wax + oxygen → carbon dioxide + water

Practical activity   Another look at combustion

We have seen that when substances burn they react with the oxygen in air.

Watch the experiment below and record your observations.

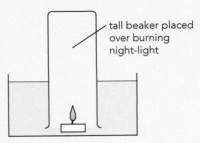

tall beaker placed over burning night-light

- What was in the upturned beaker at the start?
- What is made when wax burns?
- What is in the beaker after the reaction?
- Try to explain why the water level rises up the beaker.
  Include why this experiment does not show us how much oxygen is in the air.

In a Bunsen burner the air-hole is opened and closed by twisting its collar. The methane gas burned is mixed with more air the wider we open the air-hole. The blue flame is a 'cleaner' flame whereas the yellow flame will make a black coating of soot (carbon particles) on any object it heats. This is why the air-hole is always opened (at least slightly) to heat things up. A blue flame also releases more energy than a yellow flame. However, the yellow flame is useful when the Bunsen is lit but not used, as it is more visible than a blue flame. This is why the yellow flame with the air-hole is known as the 'safety flame'.

> ## IGCSE Link...
> You will learn more about the products of combustion in IGCSE Chemistry when you will use balanced symbol equations instead of word equations to describe the reactions.
> In symbol equations we use the chemical formulae of reactants and products.

Key terms

- **combustion**

Summary questions

1 Copy and complete:
   The chemical name for burning is _____
   When things burn they react with _____ gas in the air. The products formed are called o _____

2 Write a word equation for the combustion of wax.

3 Draw the apparatus you would use if you wanted to test which gases are given off when methane gas burns in a Bunsen burner flame.

A fire needs three things to start and keep it burning. We can show these in the '**fire triangle**':

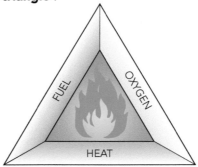

All three parts of the triangle are needed for a fire

The fuel is the store of chemical energy. When the fuel burns, its chemical energy is transferred to the surroundings as heat energy (and some light energy).

The oxygen is needed to react with the fuel in a combustion reaction. Any carbon in the fuel molecules is oxidised to carbon dioxide in a good supply of oxygen. If there is insufficient oxygen to completely oxidise the carbon, we get toxic carbon monoxide gas formed in combustion.

The heat is needed to start the fuel burning. The molecules of the fuel and the oxygen molecules have to collide with sufficient energy to cause a reaction for the fuel to burn. Once a fire starts, heat energy is released in the exothermic reaction. This supplies the energy for more and more fuel and oxygen molecules to react.

Putting out fires

If you can *remove one part of the fire triangle*, a fire goes out. For example:

- a fire blanket stops oxygen getting to the burning fuel, and so the flames die down.
- spraying water on a fire cools it down, effectively removing heat
- turning the gas off in a gas fire removes the fuel and **extinguishes** the fire.

When a fuel, heat and oxygen are all present, we get a fire

These fire fighters are spraying water on a fire to cool it down

Carbon dioxide in fire fighting

We can use carbon dioxide in fire extinguishers. The carbon dioxide from a fire extinguisher 'smothers' a fire. The gas is compressed into a metal cylinder. These extinguishers are useful on electrical and many chemical fires where water might make the fire worse. Sometimes the carbon dioxide gas is trapped in foam so that it stays on the fire. An example is in an aeroplane fire.

Another type of fire extinguisher uses a chemical reaction that produces carbon dioxide. In a soda-acid fire extinguisher, the carbon dioxide made increases the pressure in the cone-shaped extinguisher. The solution inside the cone is forced out of the nozzle and cools the fire down.

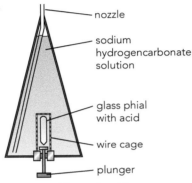

- nozzle
- sodium hydrogencarbonate solution
- glass phial with acid
- wire cage
- plunger

A soda-acid fire extinguisher

Practical activity   Research into fire-fighting

Use secondary sources to find information on how to extinguish different types of fire.

You can use reference books, DVDs, CD ROMs or the internet.

Working as a group, produce a leaflet or poster on either:

'Fire fighting' or 'Preventing fires'.

Make sure that you cover **different types of fire** and explain clearly the science involved.

Key terms

- **extinguish**
- **fire triangle**

Summary questions

1 Draw a picture to show the fire triangle.

2 In what ways can the fire triangle be applied to putting out a fire?

3 Why is it a good idea to use foam on an aeroplane fire on a runway?

4 Look at the soda-acid fire extinguisher above.
   a) Explain the way in which the soda-acid extinguisher works.
   b) Write a word equation for the reaction that takes place in a soda-acid fire extinguisher. Assume the glass phial contains dilute sulfuric acid. (Hint: hydrogencarbonates react like carbonates with dilute acid.)

Methane is the main gas in the fossil fuel called natural gas

Complete combustion

Natural gas contains mainly the compound methane, CH_4. Methane is called a **hydrocarbon** – it is a compound made up of only hydrogen and carbon atoms.

Lots of fuels are hydrocarbons. Some of the energy released when fuels burn can be transferred into useful energy.

When a hydrocarbon burns in lots of air (so that it gets enough oxygen to burn completely), we get carbon dioxide and water formed. For example, propane gas, C_3H_8, from crude oil is used in some household gas heaters:

propane + oxygen → carbon dioxide + water

Incomplete combustion

Hydrocarbon fuels can produce some black smoke as they burn. The smoke is made up of small particles of solid carbon from the fuel. Not all the carbon in the fuel is converted completely into carbon dioxide. We call this **incomplete combustion**.

In a car engine, petrol or diesel is ignited in a small space. There is not much oxygen inside the engine for the fuel to react with. So we get incomplete combustion of the fuel.

As well as carbon dioxide and water vapour, we also get:

- a toxic gas, carbon monoxide, CO
- unburnt hydrocarbon fuel
- carbon particles.

Carbon monoxide gas is so dangerous because it is odourless and effectively starves your cells of oxygen.

Cars release a variety of pollutants into the air. Catalytic converters (once they are warmed up) inside exhausts can help. They can turn carbon monoxide into carbon dioxide, nitrogen oxides into nitrogen, and unburnt hydrocarbons into carbon dioxide and water vapour.

Practical activity | Investigating burning

Your task is to find out: How does the volume of air affect the time a night-light burns? Use the apparatus shown below:

You can use a variety of different sized beakers in your tests.

Think about these points as you plan and write-up your investigation.

- Make a prediction.
  What would you expect a graph of your results to look like?
  Sketch a rough graph.

- Describe a way to measure the volume of air in each beaker?
  (Hint: You might need water, a funnel and a measuring cylinder.)

- In what ways will you make your results as accurate as possible and reduce error? Will you need to repeat your readings?

- In what way will you record your results?
  Your table should be able to show any repeat readings and averages (means) calculated.

- Plot your results on a graph.
  Will it be a bar chart or a line graph? Explain why.

- What will you do if you identify any anomalous results?

- What pattern does your graph show?

- Explain your results.
  Make sure you use your scientific knowledge and understanding.

- Evaluate your investigation.
  Suggest and explain ways in which your investigation could be improved.

Key terms

- **hydrocarbon**
- **incomplete combustion**

Summary questions

1 Write a word equation for the combustion of methane gas.

2 Explain the differences and similarities between complete combustion and incomplete combustion of a fuel, such as propane.

3 Why should a gas boiler always have a good supply of air to burn the gas?

Looking at fuels

Burning matches

Have you ever wondered how matches ignite when you strike them?

Look at the match head below:

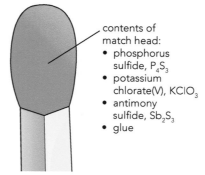

contents of match head:
- phosphorus sulfide, P_4S_3
- potassium chlorate(V), $KClO_3$
- antimony sulfide, Sb_2S_3
- glue

Matches use wood as their fuel – the head is used to ignite the wood

- When you strike the match on a rough surface, the friction causes heat energy to start a chemical reaction in the head of the match.
- The phosphorus sulfide decomposes and starts to burn. This sets off the combustion reaction of antimony sulfide.
- This reaction is made more vigorous by potassium chlorate(V), which provides extra oxygen for the combustion.
- The temperature is now hot enough to set fire to the wooden matchstick.

Comparing different fuels

One of the main factors to think about when choosing a fuel is its energy content. To compare the energy content of different fuels we need to look at the energy released per gram of fuel burned. To get this information by experiment, you need to carry out a fair test. The only variable that should change each time we collect data is the type of fuel.

One way to compare the energy released is to heat the same volume of water with each fuel in turn. You could choose to heat the water up by the same amount, for example by 10 °C, with each fuel. You will also need to weigh how much fuel is burned in each test.

To make the investigation valid, we want the energy released as the fuel burns to heat up the water. (See 'Comparing fuels' investigation below). Therefore it is best to use a copper beaker (called a calorimeter) to hold the water because copper is a good conductor of heat. However, if a copper container is not available we can use a tin can or heat-proof glassware. The important thing is to use the same material for the water's container and to position each fuel in the same place beneath it.

Practical activity   Comparing fuels

Alcohols can be used as fuels. You can compare the heat content of different alcohols using this apparatus:

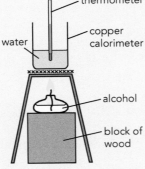

⚠ Eye protection must be worn.

Apparatus for comparing fuels

Plan a fair test.

Remember to weigh your spirit burner and its fuel before and after heating the water. This will give you the mass of fuel burned in the test.

To compare the energy content of each fuel, we can order the fuels by comparing the mass needed to heat the water by a set amount (e.g. by 10 °C). The fuel that uses up least mass has the highest energy content per gram.

Or you can put your data into the equation below to find the temperature rise per gram of fuel burned:

$$\text{Temperature rise per gram of fuel} = \frac{\text{Temperature rise in experiment}}{\text{Mass of fuel burned}}$$

- Put the fuels in order of their energy content (the one with the highest energy content first).
- What are the main sources of error in your investigation?
- What could you do to improve the validity of your data?

Summary questions

1 Look at the section on the previous page on 'Burning matches', then answer the questions:
   a) How many atoms are in a molecule of phosphorus sulfide?
   b) How many different types of atoms are there in antimony sulfide?
   c) Potassium chlorate(V) decomposes on heating to form potassium chloride and oxygen. Write a word equation for this thermal decomposition reaction.

2 An investigation was carried out to compare the energy released by three different fuels. The results were as follows:

	Mass of fuel burned / g	Volume of water heated up / cm³	Temperature change / °C
Fuel **A**	0.24	100	10
Fuel **B**	0.18	100	10
Fuel **C**	0.27	100	10

Put the fuels in order of their energy content per gram (the one with the highest energy content first).

Learning outcomes

After this topic you should be able to:

- use secondary sources to find our more about fuels
- plan investigations into dissolving ammonium chloride to test predictions made.

When choosing a fuel for a particular job, you should consider its:

- energy content
- availability/cost
- toxicity
- pollution effects
- ease of use/storage.

Practical activity   Choosing a fuel

Different fuels are needed in different situations.

- Find out which fuels are used in the following and explain why that particular fuel is chosen:
  - long-distance lorry
  - aeroplane
  - camping stove
  - rocket to launch satellites.

Endothermic investigation

As you saw on page 46, some reactions take in heat energy from their surroundings. These are called **endothermic** reactions.

Thermal decomposition is an example of an endothermic reaction. It needs continuous heating.

Practical activity   An endothermic change

 Eye protection must be worn.

Add three spatulas of ammonium chloride to 25 cm³ of water in a small beaker.

Stir it with a temperature sensor or gently with a thermometer.

Hold the beaker in the palm of your hand.

- What do you feel?
- Is heat energy given out to your hand? Or is energy being taken from your hand?

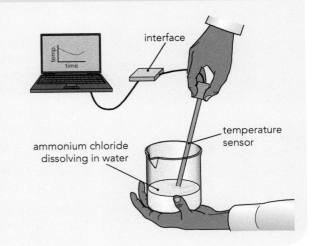

When ammonium chloride dissolves it takes in heat energy from its surroundings. Its surroundings include the beaker, the glass rod, the water, the air around it, and your hand!

Practical activity   Investigating an endothermic change

The dissolving of ammonium chloride in water is an endothermic change.

Predict what will happen to the amount of energy absorbed from the surroundings if:

a the **volume of water** is varied

b the **mass of ammonium chloride** is changed.

Plan investigations to check if your predictions are correct.

 Let your teacher look at your plan before you start your practical work.

Summary questions

1 List the factors you should consider when choosing a fuel.

2 What can you do to distinguish between an exothermic and an endothermic reaction?

3 Explain why the temperature of 100 cm³ of water decreases less than the temperature of 50 cm³ of water when an equal mass of ammonium chloride is stirred in.

Respiration releases energy

We all need the energy we get from chemical reactions. Energy heats our homes, as well as powering our appliances and the industries upon which we depend. However, more importantly, energy produced by chemical reactions in our body cells keeps us alive. The study of chemical reactions in living things is a major part of the science called **biochemistry**.

The process by which cells release some of the chemical energy stored in carbohydrate molecules is called **respiration**. This is not achieved in a single chemical reaction but in a complex series of reactions.

However, we can summarise the process of respiration by the following word equation:

glucose + oxygen → carbon dioxide + water (energy released)

The energy released in this exothermic reaction is used by the body to move, keep warm and carry out other essential functions.

If there is no oxygen available, cells can still get energy from their carbohydrates by anaerobic respiration (anaerobic means 'without oxygen'). From the same amount of glucose, this process does not release as much energy as aerobic respiration (aerobic means 'with oxygen'). There are two types of anaerobic respiration.

In the absence of oxygen, the cells in yeast survive by respiring using anaerobic respiration. They release energy from glucose and in the process make ethanol (an alcohol) and carbon dioxide.

enzymes in yeast

glucose → ethanol (alcohol) + carbon dioxide (energy released)

The other type of anaerobic respiration takes place during exercise if muscle cells cannot get enough oxygen for aerobic respiration. In this case, the glucose, in the absence of oxygen, is converted into lactic acid:

glucose → lactic acid (energy released)

The lactic acid builds up in your muscles, causing the uncomfortable 'burning' sensation you get during hard exercise. As with fermentation, the energy released is much less than in aerobic respiration.

Photosynthesis takes in energy

Plants take in carbon dioxide from the air and water from the soil. The carbon dioxide (CO_2) and water (H_2O) are the starting materials for **photosynthesis**. In a series of reactions inside the plant the carbon dioxide and water are turned into glucose ($C_6H_{12}O_6$) and oxygen (O_2) gas.

Chlorophyll, the green substance in chloroplasts inside plant cells, is needed for photosynthesis to take place. The chlorophyll absorbs light energy from the Sun which is converted into chemical energy in the products of photosynthesis.

We can summarise photosynthesis by this equation:

chlorophyll

(energy taken in) **carbon dioxide + water → glucose + oxygen**

to 'trap' light energy

This is an endothermic reaction – the energy needed to change the reactants to products is usually supplied by the Sun.

Expert tips

Note that the overall equation for photosynthesis is the reverse of the equation for respiration.

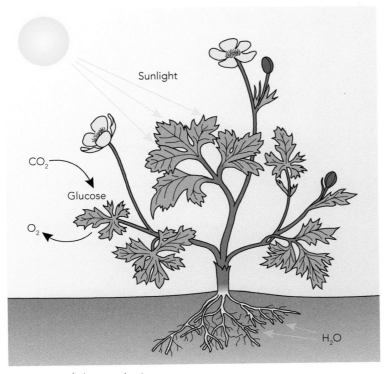

A summary of photosynthesis

The glucose made is used in the process of respiration to release energy that the plant can use. It is also used to make new substances in the plant. Much of the chemical energy in the glucose made from photosynthesis is stored in the plant as starch.

Key terms

- **biochemistry**
- **photosynthesis**
- **respiration**

Summary questions

1 What is the difference between aerobic and anaerobic respiration in terms of:
   a) oxygen present?
   b) energy released?

2 Write a word equation to summarise the process of aerobic respiration.

3 Write a word equation to summarise the process of photosynthesis.

4 Describe the role of chlorophyll in photosynthesis.

1 a What are the three parts of the fire
    triangle? *[3]*

   b i Name a suitable fire extinguisher for use
      on a fire in an electrical appliance. *[1]*

     ii Which substance should never be used
       on an electrical fire? *[1]*

   c A damp tea-towel is placed over a pan of
    burning fat to extinguish the fire. Which part
    of the fire triangle has been removed? *[1]*

   d A fire starts caused by a leaking gas pipe.
    Which part of the fire triangle should be
    removed first when fighting this fire? *[1]*

2 A student mixed the same volume of two
solutions in a polystyrene beaker. He took
the temperature of each solution and of the
mixture. Here are his results:

Solutions before mixing = 19 °C

Solution after mixing = 33 °C

   a Calculate the temperature change. *[1]*

   b Explain the results of the experiment. *[2]*

3 A group of students investigated the relative
energy content of three liquid fuels using the
apparatus below:

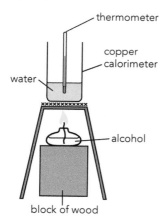

These results were obtained:

	Mass of fuel burned / g	Volume of water heated by fuel / cm³	Temperature change / °C
Fuel 1	0.2	100	10
Fuel 2	0.1	100	10
Fuel 3	0.1	200	10

   a Put the fuels in order of their energy
    content per gram (with the highest energy
    content first). *[1]*

   b Explain your reasoning to arrive at the order
    in part **a**. *[2]*

4 Copy and complete the following word
equations:

   a copper + _____ → copper oxide *[1]*

   b sodium + oxygen → _____ *[1]*

   c _____ + oxygen → water (_____ oxide) *[2]*

   d _____ + oxygen → carbon dioxide *[1]*

5 A candle is burned under a gas jar in a sand
tray:

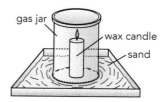

   a i When the candle burns a chemical
     reaction takes place.
     Which gas is the candle wax reacting
     with? *[1]*

     ii What is the name of this type of
       chemical reaction? *[1]*

   b As the candle burns, wax gets used up.
    Name two other ways in which we can tell
    that a chemical reaction takes place. *[2]*

   c Explain your observations in the experiment
    above. *[2]*

   d What safety precaution has been taken in
    the experiment? Why? *[2]*

   e Wax is made up of carbon and hydrogen.
    Copy and complete the word equation
    when wax burns in a plentiful supply of air:
    wax + _____ → _____ + _____ *[3]*

6 Here is a key you can use to identify gases:

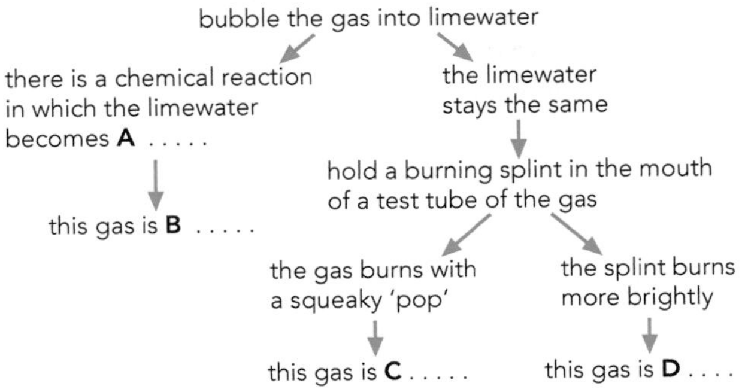

bubble the gas into limewater

there is a chemical reaction in which the limewater becomes **A** . . . . .

the limewater stays the same

this gas is **B** . . . . .

hold a burning splint in the mouth of a test tube of the gas

the gas burns with a squeaky 'pop'

the splint burns more brightly

this gas is **C** . . . . .

this gas is **D** . . . .

a What are the missing words, A to D? Choose from this list:

hydrogen

oxygen

carbon dioxide

cloudy [4]

b i What would you see if you burned a piece of magnesium ribbon in gas D? [1]

ii What safety precautions would you (or your teacher) take when doing this experiment? [2]

iii Write a word equation for the reaction in **b** part **i**. [2]

c Name the substance made when gas **C** is burned. [1]

d What is the approximate percentage of gas **D** in the air? [1]

e Which important process in a plant uses gas **B** as one of its starting materials? [1]

f Describe a test that can be used to identify the presence of water. [2]

7 One of the compounds in petrol is called octane. Its chemical formula is C_8H_{18}.

a What do we call a compound that contains hydrogen and carbon only? [1]

b Write the word equation for the burning of octane in plenty of oxygen. [2]

c Methane (CH_4) is another fuel.

i Write the word equation for the complete combustion of methane. [1]

ii Describe the results of the tests you could carry out to positively identify each product of the combustion reaction. [4]

iii Explain why a gas heater with a poor supply of oxygen due to a blocked vent can cause a fatal accident. [3]

8 a i Name the process that provides energy in our body cells. [1]

ii The process in part **a (i)** is an exothermic process. What does this mean? [1]

iii Write a word equation to describe this process. [2]

b i Which process in plants enables them to make their own food? [1]

ii The process described in **b (i)** is an endothermic process. Where do plants get the energy required for the process to take place? [1]

iii Write a word equation to describe this process in plants. [2]

c What can you say about the amount of energy involved in the two processes described in this question? Explain your answer. [2]

Glossary

A

Atomic number The number of protons (which equals the number of electrons) in an atom.

B

Biochemistry The study of the chemical reactions of biological systems.

C

Carbonate A metal compound containing the CO_3 grouping.

Catalyst A substance that speeds up a chemical reaction but remains chemically unchanged itself at the end of the reaction.

Collision theory The theory used to explain chemical reactions in terms of reacting particles colliding with sufficient energy to cause a reaction.

Combustion A chemical reaction in which a substance burns by reacting with oxygen, forming oxides and releasing energy in the process.

Concentration A measure of the quantity of a substance dissolved in a litre of solution.

Control variables The variables that could affect the outcome of an investigation; so must be kept constant for a fair test to be carried out.

D

Dependent variable The variable measured in order to judge the effect of changing the independent variable in an investigation.

Displacement reaction A chemical reaction in which a more reactive metal will displace a less reactive metal from a solution of one of its salts.

E

Electron The tiny negatively charged particles found orbiting the nucleus of atoms.

Endothermic A chemical reaction that takes in energy from its surroundings.

Energy level (shell) The area around the nucleus of an atom in which electrons are found.

Exothermic A chemical reaction that releases energy to its surroundings.

Extinguish To put out (e.g. a fire).

F

Fire triangle A way of representing the three things (fuel, heat and oxygen) we need to start or maintain a fire as the sides of a triangle.

H

Hydrocarbon A compound made up of molecules containing only hydrogen and carbon atoms.

I

Incomplete combustion When a fuel burns in a limited supply of oxygen, so oxidation of the fuel produces the pollutants carbon monoxide gas and tiny particles of unburnt carbon.

Independent variable The variable that we choose to change in an investigation.

N

Neutron Neutral sub-atomic particles, with the same mass as a proton, found in the nucleus of an atom.

Nucleus The centre of the atom, containing protons and neutrons.

O

Ore Rock from which it is economically worthwhile to extract metal.

P

Periodic Table The table of chemical elements listed in order of their atomic number, with eight main groups of elements. Within each group, the elements often have similar properties.

Photosynthesis The process in which plants make their own food from carbon dioxide and water reacting in the presence of light to form glucose and oxygen.

Proton Dense, positively charged sub-atomic particles found in the nucleus of atoms.

Q

Qualitative data Data that is collected by observation.

Quantitative data Data that is collected by measurements.

R

Reactivity Series A list of mainly metallic elements placed in order of their reactivity.

Respiration The process in which cells release the energy stored in glucose molecules from its reaction with oxygen, forming the waste products carbon dioxide and water.

T

Tarnish When the surface of a metal loses its shiny appearance due to oxidation reactions with gases (mainly oxygen) in the air.

Trend Regular pattern in a particular direction.

V

Variables Factors that can change in an investigation (see independent, dependent and control variables).

Index

Note: Key (glossary) terms are in **bold** type.

Index

Photo acknowledgements

Alamy: Studiomode, p44 (top right); David R.Frazier Photolibrary, Inc, p44 (bottom right); **Fotolia:** p14; p50 (left); **iStockphoto:** p2; p22; p25; p32; p34; p38; p43; p44 (top left); p45; p49; p50 (right); p52; p56; **Martyn Chillmaid:** p10; p24; **Science Photo Library:** Adam Hart-Davies, p3; Andrew Lambert Photography, p11; Pascal Goetgheluck, p23; Andrew Lambert Photography, p28; p41; Martyn F. Chillmaid, p44 (bottom left).